"十三五"职业教育国家规划教材
"十二五"职业教育国家规划教材

高职高专电子信息类"十三五"规划教材

电子与通信技术
专业英语（第5版）

刘骋 蔡静 刘小芹 主编

人民邮电出版社
北京

图书在版编目（CIP）数据

电子与通信技术专业英语 / 刘骋，蔡静，刘小芹主编. -- 5版. -- 北京：人民邮电出版社，2019.6（2021.12重印）
高职高专电子信息类"十三五"规划教材
ISBN 978-7-115-51145-4

Ⅰ. ①电… Ⅱ. ①刘… ②蔡… ③刘… Ⅲ. ①电子技术－英语－高等职业教育－教材②通信技术－英语－高等职业教育－教材 Ⅳ. ①TN-43

中国版本图书馆CIP数据核字(2019)第075913号

内 容 提 要

本书是一本工学结合的专业英语教材，共分6个单元。其中前4个单元共24课，主要包括电子技术基础知识、电子仪器和设备、通信技术、电子高新技术方面的内容。每课内容包括课文、专业词汇表、注解和练习，学习重点放在阅读理解、专业词汇积累和书面翻译上。后2个单元为专业英语基础知识部分，主要介绍专业英语阅读（翻译）及写作基础知识。全书参考教学课时为60学时。

本书主要作为高等职业院校、高等专科院校电子与通信类专业的教学用书，还可作为成人民办高校及相关人员的教学与学习参考书或专业英汉对照培训用书。

◆ 主　编　刘　骋　蔡　静　刘小芹
　责任编辑　左仲海
　责任印制　马振武

◆ 人民邮电出版社出版发行　北京市丰台区成寿寺路11号
　邮编　100164　电子邮件　315@ptpress.com.cn
　网址　http://www.ptpress.com.cn
　固安县铭成印刷有限公司印刷

◆ 开本：787×1092　1/16
　印张：14.75　　　　　　2019年6月第5版
　字数：377千字　　　　　2021年12月河北第8次印刷

定价：49.80元

读者服务热线：(010)81055256　印装质量热线：(010)81055316
反盗版热线：(010)81055315
广告经营许可证：京东市监广登字20170147号

前言

　　本书是由院校教师和电信行业技术人员联合编写的一本工学结合的专业英语教材,编者立足科学技术发展前沿,不断调整和更新教学内容,及时将当下电子与通信行业、企业的新科学、新技术、新工艺呈现给在校大学生。新版取材于国内外工程资料,全方位紧密结合专业知识,按照企业电子与通信技术人员的典型工作任务的流程和学生学习知识与技能的认知过程编排,可适应电子与通信技术不断更新的新情况,并可满足高等职业教育改革中工学结合、任务引领的需要。

　　本次修订新增了两个单元,分别介绍了专业英语阅读(翻译)知识与写作知识,可提高学生专业英语的阅读(翻译)与写作水平。新编教材体现了人才培养的层次性、知识结构的交融性和教学内容的实践性。

　　本书的特点主要体现在以下几个方面。

　　(1)各章均采用问题导入式教学方法,部分课文采用案例式教学方法,让学生带着工作具体任务学习。力求做到学习内容的宽度和深度循序渐进,尽量简化长难句,图文并茂,让学生在较短时间内熟悉专业文章、工程资料和操作手册的英文表达,并积累一定数量的专业词汇,使学生能够更直观地了解所学内容与实物的联系,培养阅读和应用电子与通信技术原版资料的能力。

　　(2)本书既保留了基础的电工电子专业知识,以及无线电波、电路图和方框图等基础知识,又介绍了现代通信、计算机和电子技术等新知识,以及近年来推广的光纤通信、移动通信、卫星通信、宽带通信、多媒体信息服务、卫星电话、ISDN 技术、全球定位系统等先进技术。本书还加入了计算机视觉、计算机仿真、多媒体技术、人工智能、数字图像处理、人工神经网络、虚拟仪器、Wi-Fi、智能手机、物联网、遥感技术等电子与通信技术的相关内容,不仅强化了学生的英语水平,而且新增了工程实践知识,对于开阔学生的视野,帮助学生了解行业动态、培养学习兴趣起到了关键作用。

　　(3)本书所选的英文材料全部来自英文原版资料,用词、句型、语法结构全部遵循英文使用习惯,有利于培养学生使用准确的英语表达方式的习惯,避免养成"自创"英语的不良习惯。

　　(4)本书配有习题及多媒体课件,由来自以英语为母语的国家的外籍教师录制音频资料,供学生练习口语和正音,帮助学生大胆开口,强化专业英语听、说的能力。学生可利用课外时间随时练习,既能节省课堂时间,又可反复练习,为对外业务交流打下一定的基础。总之,根据语言学习的特点,本书力求以学生能应用所学知识为重点,在专业英语的教学上也力求做到培养应用型人才。

　　本书的 Unit Ⅲ、Unit Ⅳ和 APPENDIX Ⅰ由刘骋编写,Unit Ⅱ、Unit Ⅴ、Unit Ⅵ由蔡静编写,Unit Ⅰ和 APPENDIX Ⅱ由刘小芹编写,全书由刘骋统稿审校。来自企业的刘文、张敬衡、杨新明、胡柏利、罗中平、唐小琦、蒋开勤、周凌等技术专家,以及来自澳大利亚的机电工程专家 Mr. Bruce Skewes 对本书进行了指导和帮助。本书获评"十二五"职业教育国家规划教材,相关专家学者也对本书的修订给予了高度认可，并提出了宝贵的修订建议，本书编者团队在此一并致谢！

  由于编者水平有限，时间仓促，加上形势的发展也在不断提出新的要求，书中难免有不足和欠妥之处，敬请读者批评指正。

编者

2018 年 11 月

目 录

第1单元 电子技术基础知识 1

    第1课　认识电子元件 7

    第2课　电流、电压和电阻 11

    第3课　交流电、直流电及电信号 15

    第4课　晶体管电压放大器 20

    第5课　数字电路 25

    第6课　电路图和方框图 29

    阅读材料 31

        1. 基本电阻电路 31

        2. 电容和电感 32

        3. 二极管及其电路 33

        4. 晶体管及其基本电路 34

        5. 无线电波 36

        6. 电源 38

第2单元 电子仪器和设备 39

    第7课　万用表及其使用 42

    第8课　示波器 48

    第9课　虚拟仪器 53

    第10课　便携式媒体播放器 57

    第11课　液晶显示器 63

    第12课　智能手机 69

    阅读材料 71

        7. 数字电压表 71

        8. 触摸屏 72

        9. 高清电视 74

        10. 数字音响 75

        11. 家庭影院 77

        12. 数字机顶盒 78

第 3 单元　通信技术 80

第 13 课　光纤通信 84
第 14 课　卫星通信 88
第 15 课　无线相容性认证 94
第 16 课　全球定位系统 99
第 17 课　无线传感器网络 104
第 18 课　4G 网络 108
阅读材料 110
　　13. 移动通信 110
　　14. 宽带通信 111
　　15. 路由器 113
　　16. 有线电视 114
　　17. 交互式网络电视 115
　　18. 三网合一 117

第 4 单元　电子高新技术 119

第 19 课　物联网 123
第 20 课　计算机仿真 129
第 21 课　遥感技术 133
第 22 课　人工智能 139
第 23 课　二维码 143
第 24 课　光伏技术 147
阅读材料 149
　　19. 数字图像处理 149
　　20. 计算机视觉 150
　　21. 多媒体技术 152
　　22. 人工神经网络及其应用 153
　　23. 射频识别 154
　　24. 智慧城市 156

第 5 单元　专业英语基础知识 I 158

第 6 单元　专业英语基础知识 II 191

附录 I　词汇表 213

附录 II　常用缩写 226

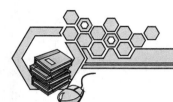

Contents

Unit I Basic Knowledge of Electronics 1

Lesson 1   Knowing the Electronic Components 1
Lesson 2   Current, Voltage and Resistance 8
Lesson 3   AC, DC and Electrical Signals 12
Lesson 4   Transistor Voltage Amplifier 17
Lesson 5   Digital Circuit 22
Lesson 6   Circuit Diagrams and Block Diagrams 26
Reading Material 31
    1. Basic Resistor Circuits 31
    2. Capacitance and Inductance 32
    3. Diodes and Its Circuit 33
    4. The Transistor and Its Basic Circuit 34
    5. Radio Waves 36
    6. Power Supply 38

Unit II Electronic Instruments and Products 39

Lesson 7   Multimeter and Its Usage 39
Lesson 8   The Oscilloscope 44
Lesson 9   Virtual Instruments 49
Lesson 10   Portable Media Player 54
Lesson 11   LCD (Liquid Crystal Display) 59
Lesson 12   Smartphone 65
Reading Material 71
    7. Digital Voltmeter 71
    8. Touchscreen 72
    9. High Definition TV (HDTV) 74
    10. Digital Audio 75
    11. Home Theater System 77
    12. Set-Top Box (STB) 78

Unit Ⅲ　Communicated Technology 80

Lesson 13　Optical Fiber Communications 80
Lesson 14　Satellite Communications 85
Lesson 15　Wireless Fidelity 90
Lesson 16　Global Positioning System (GPS) 95
Lesson 17　Wireless Sensor Network 100
Lesson 18　4G Network 105
Reading Material 110
　　13. Mobile Communications 110
　　14. Broadband Communications 111
　　15. Router 113
　　16. Cable Television 114
　　17. Internet Protocol Television (IPTV) 115
　　18. Integration of Three Networks 117

Unit Ⅳ　Advanced Electronic Technology 119

Lesson 19　Internet of Things 119
Lesson 20　Computer Simulation 125
Lesson 21　Remote Sensing 130
Lesson 22　Artificial Intelligence (AI) 135
Lesson 23　2D Bar Code 140
Lesson 24　Photovoltaic Technology 144
Reading Material 149
　　19. Digital Image Processing 149
　　20. Computer Vision 150
　　21. Multimedia Technology 152
　　22. Artificial Neural Networks and Their Applications 153
　　23. Radio Frequency Identification (RFID) 154
　　24. Intelligent City 156

Unit Ⅴ　Basic Knowledge of Professional English Ⅰ 158

Unit Ⅵ　Basic Knowledge of Professional English Ⅱ 191

APPENDIX Ⅰ　Vocabulary 213

APPENDIX Ⅱ　Abbreviations 226

Unit I

Basic Knowledge of Electronics

电子技术基础知识

本章的内容全部是学生熟悉的电子技术的基础知识,包括对电子元器件的一般介绍,电流和电压的引入,对电阻、电容、电感、二极管和三极管的具体分析,对直流电和交流电、电路图和方框图的介绍等。目的是将学生过去已有的普通英语和专业技术基础知识有机地结合,在理解的基础上掌握工程英语常用表达方式、典型句型和专业术语,建立工程英语概念,为深入学习电子信息技术专业英语打下基础。

Lesson 1 Knowing the Electronic Components

There are a large number of symbols which represent an equally large range of electronic components. It is important that you can recognize the more common components and understand what they actually do. A number of these components are drawn below, and it is interesting to note that often there is more than one symbol representing the same type of component.

Resistors

Resistors restrict the flow of electric current, for example, a resistor is placed in series with a Light-Emitting Diode (LED) to limit the current passing through the LED. Fig.1-1 shows resistor example and circuit symbol.

Fig.1-1　Resistor Example and Circuit Symbol

Resistors may be connected either way round. They are not damaged by heat when soldering.

Capacitors

Capacitors store electric charge. They are often used in filter circuits because capacitors easily pass AC (changing) signals but they block DC (constant) signals. Fig.1-2 shows capacitor examples and circuit symbol.

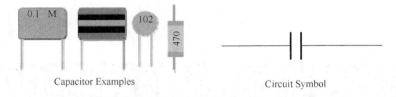

    Capacitor Examples   Circuit Symbol

Fig.1-2 Capacitor Examples and Circuit Symbol

Inductor

An inductor is a passive electronic component that stores energy in the form of a magnetic field. An inductor is a coil of wire with many windings, often wound around a core made of a magnetic material, like iron. Fig.1-3 shows inductor examples and circuit symbol.

    Inductor Examples   Circuit Symbol

Fig.1-3 Inductor Examples and Circuit Symbol

Diodes

Diodes allow electricity to flow in only one direction. The arrow of the circuit symbol shows the direction in which the current can flow. Diodes are the electrical version of a valve and early diodes were actually called valves. Fig.1-4 shows diode examples and circuit symbol.

    Diode Examples   Circuit Symbol

Fig.1-4 Diode Examples and Circuit Symbol

Transistors

There are two types of standard transistors, **NPN** and **PNP**, with different circuit symbols. The letters refer to the layers of semiconductor material used to make the transistor. Fig.1-5 shows transistor examples and circuit symbols.

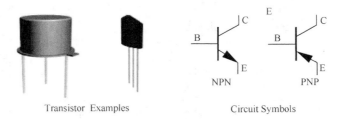

Transistor Examples   Circuit Symbols

Fig.1-5   Transistor Examples and Circuit Symbols

Integrated Circuits (Chips)

Integrated Circuits are usually called ICs or Chips. They are complex circuits which have been etched onto tiny chips of semiconductor (silicon). The chip is packaged in a plastic holder with pins spaced on a 0.1"(2.54 mm) grid which will fit the holes on stripboards and breadboards. Very fine wires inside the package link the chip to the pins. Fig.1-6 shows integrated circuits example.

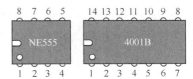

Fig.1-6   Integrated Circuits Example

Light Emitting Diodes (LEDs)

LEDs emit light when an electric current passes through them.

LEDs must be connected the correct way round, and the diagram may be labelled "**a**" or "+" for anode and "**k**" or "−" for cathode (yes, it really is "k", not "c", for cathode!). The cathode is the short lead and there may be a slight flat on the body of round LEDs. Fig.1-7 shows LED example and circuit symbol.

LED Example   Circuit Symbol

Fig.1-7   LED Example and Circuit Symbol

Other Electronic Components

Fig.1-8 shows other electronic component examples and circuit symbols.

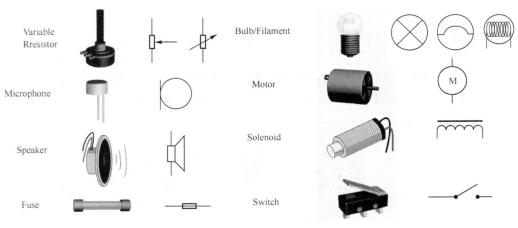

Fig.1-8　Other Electronic Component Examples and Circuit Symbols

New Words

electronic [iˌlek'trɔnik] adj.　　电子的，电子仪器的
electronics [iˌlek'trɔniks] n.　　电子学，电学，电子工业
component [kəm'pəunənt] n. & adj.　　成分，元件；组成的，构成的
symbol ['simbl] n.　　符号，标志，象征
resistor [ri'zistə] n.　　电阻器
restrict [ri'strikt] vt.　　限制，约束，限定
current ['kʌrənt] n.　　电流
series ['siəri:z] n.　　连续，系列，丛书，级数
capacitor [kə'pæsitə] n.　　电容器
charge [tʃɑ:dʒ] n.　　电荷
inductor [in'dʌktə] n.　　感应器，电感
magnetic [mæg'netik] adj.　　磁的，有磁性的，有吸引力的
field ['fi:ld] n.　　场
diode ['daiəud] n.　　二极管
valve [vælv] n.　　电子管，真空管
transistor [træn'sistə] n.　　晶体管
integrated ['intigreitid] adj.　　综合的，完整的
circuit ['sə:kit] n.　　电路
chip [tʃip] n.　　芯片
semiconductor [ˌsemikən'dɔktə] n.　　半导体
silicon ['silikən] n.　　硅，硅元素
anode ['ænəud] n.　　阳极，正极
cathode ['kæθəud] n.　　阴极

microphone ['maikrəfəun] n.	扩音器，麦克风
speaker ['spi:kə] n.	扬声器，喇叭
fuse [fju:z] n.	保险丝，熔丝
filament ['filəmənt] n.	细丝，灯丝
motor ['məutə] n.	发动机，电动机
solenoid ['səulənɔid] n.	螺线管
switch [switʃ] n. & vt.	开关；转换

Phrases and Expressions

light-emitting diode	发光二极管
AC (Alternating Current)	交流电
DC (Direct Current)	直流电

Notes

1. There are a large number of symbols which represent an equally large range of electronic components.

   译文：这些符号代表了绝大多数的电子元件。

2. It is important that you can recognize the more common components and understand what they actually do.

   译文：能够识别更多的普通元件，以及了解它们的实际用途是很重要的。

3. A number of these components are drawn below, and it is interesting to note that often there is more than one symbol representing the same type of component.

   译文：下面画出了许多电子元件，有趣的是，一种类型的元件常常可以用多种符号表示。

4. They are complex circuits which have been etched onto tiny chips of semiconductor (silicon). The chip is packaged in a plastic holder with pins spaced on a 0.1"(2.54 mm) grid which will fit the holes on stripboards and breadboards.

   译文：它们是固化在微小的半导体（硅）芯片上的复杂电路。该芯片被封装在一个塑料固定物上，引脚间距为0.1英寸（2.54 mm），适合条形板和面包板的孔距。

Exercises

1. Write T (True) or F (False) beside the following statements about the text.

a. One symbol represents a type of component.

b. Resistors are damaged by heat when soldering.

c. Capacitors are not often used in filter circuits.

d. Both capacitor and inductor are passive electronic component.

e. Inductor stores electric charge.

f. Diodes allow electricity to flow either one direction.

g. Diodes are also called valves.

h. The symbol of NPN and PNP transistors is same.

i. ICs are complex circuits which have been etched onto tiny chips of semiconductor.

j. LEDs is one type of diodes.

2. Match the following terms to appropriate definition or expression.

a. AC　　　　　　　　1. A tiny "chip" containing many individual circuits which work together to perform a function.

b. DC　　　　　　　　2. A material that is neither a conductor nor an insulator.

c. semiconductor　　　3. The direction of current is constant.

d. IC　　　　　　　　4. A portion of space surrounding a body in which the forces due to the body can be detected.

e. field　　　　　　　5. It changes polarity periodically.

3. Fill in the missing words according to the text.

a. In the common components, _____ may be connected either way round.

b. _____ is a passive electronic component that stores electric charge.

c. Inductors store energy in the form of _____.

d. Diodes allow electricity to flow in _____ direction.

e. The _____ of the LEDs is the short lead and there may be a slight flat on the body of round LEDs.

4. Translate the following sentences into Chinese.

a. Modern advances in the fields of computer, control system, communications have a close relationship with electronics.

b. The field of electronics includes the electron tube, transistor, integrated circuit and so on.

c. Although resistors, capacitors and inductors form important elements in electronic circuit, it is essential to know something about resistance, capacitance and inductance.

d. Electronic technology is developing rapidly in the world. And electronics industry is equipped to make yet another giant step forward.

Chinese Translation of Texts（参考译文）

第1课　认识电子元件

这些符号代表了绝大多数的电子元件。能够识别更多的普通元件，以及了解它们的实际用途是很重要的。下面画出了一些电子元件，有趣的是，一种类型的元件常常可以用多种符号表示。

电阻

电阻阻碍电流的流动，例如，一个电阻与一个发光二极管LED串联来限制流过LED的电流。图1-1为电阻实物图和电路符号。电阻可以连接在任一回路中。它们不会因焊接高温而损坏。

电容

电容存储电荷。因为电容使交流信号容易通过而阻隔直流信号，所以它们经常被用在滤波电路中。图1-2为电容实物图和电路符号。

电感

电感是一个无源电子元件，它以磁场的形式存储能量。电感是一个被电线缠绕了许多圈的线圈，经常缠绕在像铁这样的磁材料的磁芯上。图1-3为电感实物图和电路符号。

二极管

二极管允许电流仅从一个方向流过。电路符号的箭头指示了电流能流过的方向。二极管是真空管的电子版，实际上，早期的二极管就叫真空管。图1-4为二极管实物图和电路符号。

晶体管

标准的晶体管有两种类型，NPN型和PNP型，它们的电路符号不同。字母表示制造晶体管的半导体材料。图1-5为晶体管实物图和电路符号。

集成电路（芯片）

集成电路通常被称为IC或芯片。它们是固化在微小的半导体（硅）芯片上的复杂电路。该芯片被封装在一个塑料固定物上，引脚间距为0.1英寸（2.54 mm），适合条形板和面包板的孔距。在封装里面用非常纤细的导线连接芯片的引脚。图1-6为集成电路实物图。

发光二极管（LEDs）

当电流流过LEDs时，它发出光。

LEDs必须连接正确的回路，电路图中可以标"a"或者"+"表示阳极，标"k"或者"-"表示阴极。阴极是短的引脚并且在LEDs圆形体内可能是微小扁平的那端。图1-7为LED实物图和电路符号。

其他的电子元件

图1-8为其他电子元件的实物图和电路符号。

可变电阻器、蜂鸣器（麦克风）、扬声器、保险丝、灯泡/灯丝、电动机、螺线管、开关。

Lesson 2 Current, Voltage and Resistance

The flow of electrons through a conductor is called a current. Current flow is represented by the letter symbol I. The basic unit in which current is measured is the ampere (A). One ampere of current is defined as the movement of one coulomb (6.28×10^{18} electrons) past any point of a conductor during one second of time.

When it is desirable to express a magnitude of current smaller than the ampere, the milliampere (mA) and the microampere (μA) units are used. One milliampere is equivalent to one-thousandth (0.001) of an ampere, and one microampere is equivalent to one-millionth (0.000001) of an ampere.

The term voltage (represented by the letter symbol U) is commonly used to indicate both a difference in potential and an electromotive force. The unit in which voltage is measured is the volt. One volt is defined as that magnitude of electromotive force required cause a current of one ampere to pass through a conductor having a resistance of one ohm.

A magnitude of voltage less than one volt is expressed in terms of millivolts (mV) or microvolts (μV). Larger magnitudes of voltage are expressed in kilovolts (kV). One kilovolt equals one thousand volts.

The opposition to current is called electrical resistance and is represented by the letter symbol R. The unit of resistance is the ohm, a term that is often expressed by using Ω. One ohm is defined as that amount of resistance that will limit the current in a conductor is one ampere when the voltage applied to the conductor is one volt. Larger amounts of resistance are commonly expressed in kiloohm (kΩ) and in megohm (MΩ).

The relationship between volts, amperes, and ohms can be represented by "Ohm's Law". "Ohm's Law" states that the ratio of the voltage between the ends of a wire and the current flowing in it is equal to the resistance of the wire. Now we can say that when a given voltage is applied across the ends of the wire, an electric current always flows along it, and the value of this current depends on the resistance of wire (as shown in Fig.2-1).

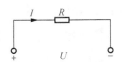

Fig.2-1 Current, Voltage and Resistance

New Words

conductor [kən'dʌktə] n.	导体，导线
ampere ['æmpɛə] n.	安培
coulomb ['ku:lɔm] n.	库仑
magnitude ['mægnitju:d] n.	大小，数量，巨大，广大
equivalent [i'kwivələnt] adj.	相等的，相同的，等量的
volt [vəult] n.	伏特

Phrases and Expressions

electromotive force　　　　　　　　电动势
electric current　　　　　　　　　　电流

Notes

1. The flow of electrons through a conductor is called a current.
   译文：通过导体的电子流称为电流。

2. The basic unit in which current is measured is the ampere (A).
   译文：度量电流的基本单位是安培。
   句中 in which current is measured 是定语从句，修饰 unit。

3. to be defined as
   "给……下定义为"，后可接名词或宾语从句，如课文中第一段。

4. When it is desirable to express a magnitude of current smaller than the ampere, the milliampere (mA) and the microampere (μA) units are used.
   译文：当需要表示比安培小的电流量时，可用毫安（mA）和微安（μA）。
   it 是形式主语，不定式 to express a magnitude of current smaller than the ampere 充当真正的主语。

5. One volt is defined as that magnitude of electromotive force required cause a current of one ampere to pass through a conductor having a resistance of one ohm.
   译文：使 1 安培电流流过电阻为 1 欧姆的导体所需的电动势定义为 1 伏特。
   （1）过去分词 required，充当后置定语，修饰 electromotive force。（2）having a resistance of one ohm 是现在分词短语，充当后置定语，修饰 conductor。

6. The unit of resistance is the ohm, a term that is often expressed by using Ω.
   译文：电阻的单位是欧姆，常用 Ω 表示。
   a term… 为 ohm 的同位语，其中 that is often expressed by using Ω 是定语从句，修饰 term。

7. One ohm is defined as that amount of resistance that will limit the current in a conductor is one ampere when the voltage applied to the conductor is one volt.
   译文：1 欧姆的定义是，当加到导体上的电压为 1 伏特时，将导体中的电流限制为 1 安培所需的电阻值。
   that 引导定语从句，修饰 amount of resistance，其中 when the voltage applied to the conductor is one volt 是时间状语从句。

8. The relationship between volts, amperes, and ohms can be represented by "Ohm's Law".
   译文：伏特、安培和欧姆之间的关系可用欧姆定律表示。
   欧姆定律，即电阻等于电压除以电流，$R = U/I$（$U = IR$ 或 $I = U/R$）。

Exercises

1. Write T (True) or F (False) beside the following statements about the text.

a. Current flow is represented by the letter symbol I.

b. One milliampere is equivalent to one-thousandth (0.001) of an ampere, and one microampere is equivalent to one-billionth (0.000000001) of an ampere.

c. The term voltage is commonly used to indicate a difference in potential but electromotive force is not.

d. One kilovolt equals one thousand volts.

e. The opposition to current is called electrical resistance.

f. Larger amounts of resistance are commonly expressed in kiloohm (kΩ) and in megohm (MΩ).

g. The flow of electrons through a conductor is called a resistance.

h. One volt is defined as that magnitude of electromotive force required cause a current of one ampere to pass through a conductor having a resistance of one ohm.

2. Match the following terms to appropriate definition or expression.

a. current　　　　　1. electromotive force

b. amp　　　　　　2. the relationship between volts, amperes, and ohms

c. voltage　　　　　3. the flow of electrons

d. ohm　　　　　　4. the unit of resistance

e. Ohm's Law　　　5. the unit in which current is measured

3. Fill in the missing words according to the text.

a. One ampere of current is defined as the movement of _____ coulomb (6.28×10^{18} electrons) past any point of a conductor during _____ second of time.

b. One milliampere is equivalent to _____ (0.001) of an ampere, and one microampere is equivalent to _____ (0.000001) of an ampere.

c. One volt is defined as that magnitude of electromotive force required cause a current of one to pass through a conductor having a _____ of one ohm.

d. One ohm is defined as that amount of _____ that will limit the current in a conductor is one ampere when the _____ applied to the conductor is one volt.

4. Translate the following paragraphs into Chinese.

Potential

The unit for potential difference, or electromotive force, is the volt. The abbreviation, or symbol, for this unit is V. Voltage is expressed in volts. Recall that one volt equals the amount of electromotive force (emf) that moves a current of one ampere through a resistance of one ohm.

Current

The unit of measure for current flow is the ampere. The abbreviation, or symbol, for this basic unit of measure is A. Remember that one ampere equals an electron flow of one coulomb per second past a given point.

Resistance

Resistance is another electrical parameter that two letter: "R" represents the general term resistance and the Greek letter omega (Ω) represents the unit of resistance, the ohm. Remember that one ohm equals the resistance that limits the current to one ampere with one volt applied.

Conductance

Another electrical parameter is conductance. Conductance is the opposite of resistance. The unit of conductance is the siemens (S) named after the scientist Ernst Siemens. The abbreviation for the general term conductance is G.

5. Translate the following sentences into English.

a. 电压的单位是伏特，用符号 V 表示。

b. 电流的单位是安培，用符号 A 表示。

c. 1 伏特的电压施加在导体上产生了 1 安培的电流，此时电阻为 1 欧姆。

d. 欧姆定律表示了电流、电压、电阻之间的关系。

Chinese Translation of Texts（参考译文）

第 2 课　电流、电压和电阻

通过导体的电子流称为电流。电流用字母符号 I 表示。度量电流的基本单位是安培。1 安培电流的定义是，在 1 秒钟内有 1 库仑（6.28×10^8 个电子）的电量通过导体的任何一点时的电流为 1 安培。

当需要表示比安培小的电流量时，可用毫安（mA）和微安（μA）。1 毫安等于 0.001 安培，1 微安等于 0.000001 安培。

电压这个术语（用字母符号 V 表示）常用来表示电位差和电动势。度量电压的单位是伏特。使 1 安培电流流过电阻为 1 欧姆的导体所需的电动势定义为 1 伏特。

小于 1 V 的电压值用毫伏（mV）或微伏（μV）表示。较高的电压值可用千伏（kV）表示。1 kV 等于 1000 V。

对电流的阻力叫作电阻，用字母符号 R 表示。电阻的单位是欧姆，常用 Ω 表示。1 欧姆的定义是，当加到导体上的电压为 1 伏特时，将导体中的电流限制为 1 安培所需的电阻值。较大的电阻值常用千欧（kΩ）和兆欧（MΩ）表示。

伏特、安培和欧姆之间的关系可用欧姆定律表示。欧姆定律描述为，导线两端电压与流过导线的电流的比是导线的电阻。现在我们可说，当在导线两端施加一给定电压时，导线中总是有电流，电流的值取决于导线的电阻（见图 2-1）。

Lesson 3 AC, DC and Electrical Signals

Alternating Current (AC)

Alternating Current (AC) flows one way, then the other way, continually reversing direction (as shown in Fig.3-1 and Fig.3-2). An AC voltage is continually changing between positive (+) and negative (−). The rate of changing direction is called the frequency of the AC, and it is measured in hertz (Hz) which is the number of forwards-backwards cycles per second.

An AC supply is suitable for powering some devices such as lamps and heaters, but almost all electronic circuits require a steady DC supply.

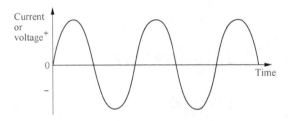

Fig.3-1 AC from a Power Supply:
This shape is called a sine wave

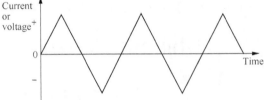

Fig.3-2 This triangular signal is AC because it
changes between positive (+) and negative (−)

Direct Current (DC)

Direct Current (DC) always flows in the same direction, but it may increase and decrease. A DC voltage is always positive (or always negative), but it may increase and decrease. Electronic circuits normally require a steady DC supply which is constant at one value (as shown in Fig.3-3). Cells, batteries and regulated power supplies provide steady DC which is ideal for electronic circuits. Lamps, heaters and motors will work with any DC supply.

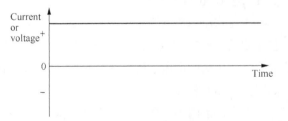

Fig.3-3 Steady DC: from a battery or regulated power
supply, this is ideal for electronic circuits

Properties of Electrical Signals

An electrical signal is a voltage or current which conveys information, usually it means a voltage. The term can be used for any voltage or current in a circuit.

The voltage-time graph on the Fig.3-4 shows various properties of an electrical signal. In addition to the properties labeled on the graph, there is frequency which is the number of cycles per second. The diagram shows a sine wave but these properties apply to any signal with a constant shape.

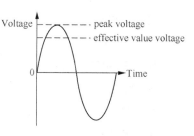

Fig.3-4　Various Properties of an Electrical Signals

Amplitude is the maximum voltage reached by the signal. It is measured in volts, V.

Peak voltage is another name for amplitude.

Peak-peak voltage is twice the peak voltage (amplitude). When reading an oscilloscope trace, it is usual to measure peak-peak voltage.

Time period is the time taken for the signal to complete one cycle. It is measured in seconds (s), but time periods tend to be short, so milliseconds (ms) and microseconds (μs) are often used. 1 ms = 0.001s and 1 μs = 0.000001 s.

Frequency is the number of cycles per second. It is measured in hertz (Hz), but frequencies tend to be high, so kilohertz (kHz) and megahertz (MHz) are often used. 1 kHz=1000 Hz and 1 MHz=1000000 Hz. Frequency=1/time period, and time period= 1/frequency.

Another value used is the effective value of AC. This is the value of alternating voltage or current that will have the same effect on a resistance as a comparable value of direct voltage or current will have on the same resistance.

New Words

positive ['pɔzətiv] adj.	正的
frequency ['fri:kwənsi] n.	频率
measure ['meʒə] v. & n.	测量
device [di'vais] n.	装置，设备
label ['leibl] n. & v.	标签；贴标签于……
amplitude ['æmplitju:d] n.	振幅
period ['piəriəd] n.	时期，周期
oscilloscope [ɔ'siləskəup] n.	示波器
millisecond ['milisekənd] n.	毫秒
microsecond ['maikrəusekənd] n.	微秒

Phrases and Expressions

Alternating Current	交流电
Direct Current	直流电

power supply	电源
electronic circuit	电子电路
electrical signal	电信号
peak voltage	峰值电压
peak-peak voltage	峰-峰值电压
effective value	有效值

Notes

1. The rate of changing direction is called the frequency of the AC, and it is measured in hertz (Hz) which is the number of forwards-backwards cycles per second.

   译文：这种方向变化的速率称为交流信号的频率，单位是赫兹，它表示一秒内周期性变化的次数。

   hertz (Hz)：n. 赫兹（电学频率单位），赫（兹），每秒周数【频率单位】，【大写】赫兹（德国物理学家）

   cycles per second：每秒循环数。

2. The voltage-time graph on the Fig.3-4 shows various properties of an electrical signal.

   译文：图 3-4 的电压-时间图显示了电信号的各种特性。

   voltage-time graph：电压时间曲线。

3. The diagram shows a sine wave but these properties apply to any signal with a constant shape.

   译文：该图显示了一个正弦波，但这些特性适用于具有恒定形状的任何信号。

   sine wave：n.正弦波。

   cosine wave：n.余弦波。

4. Time period is the time taken for signal to complete one cycle. It is measured in seconds (s), but time periods tend to be short so, milliseconds (ms) and microseconds (μs) are often used. 1 ms = 0.001 s and 1 μs = 0.000001 s.

   译文：周期是信号完成一个周期所需要的时间。周期的单位是秒，但是周期往往非常短，所以常用毫秒（ms）和微秒（μs）来表示。1 ms =0.001 s，1 μs=0.000001 s。

   milliseconds (ms)：n. 毫秒。

   microseconds (μs)：micro-second n. 微秒。

5. 1 kHz = 1000 Hz and 1 MHz = 1000000 Hz.

   译文：千赫（kHz）和兆赫（MHz）。

6. Frequency=1/time period, and time period=1/frequency.

   译文：频率=1/周期，周期=1/频率。

   time period：周期。

Unit I Basic Knowledge of Electronics 电子技术基础知识

Exercises

1. Write T (True) or F (False) beside the following statements about the text.

a. Direct Current (DC) flows one way, then the other way, continually reversing direction.

b. Alternating Current (AC) always flows in the same direction.

c. Amplitude is the minimum voltage reached by the signal.

d. Peak-peak voltage is twice the peak voltage (amplitude).

e. The effective value of AC is the value of alternating voltage or current that will have the same effect on a resistance as a comparable value of direct voltage or current will have on the same resistance.

2. Fill in the missing words according to the text.

a. An AC voltage is continually changing between _____ and _____.

b. Electronic circuits normally require a steady DC supply which is _____ one value.

c. Cells, batteries and regulated power supplies provide _____ which is ideal for electronic circuits.

d. In addition to the properties labeled on the graph, there is frequency which is the number of _____ per second.

e. When reading an oscilloscope trace, it is usual to measure _____ voltage.

3. Match the following terms to appropriate definition or expression.

a. an electrical signal   1. 1/time period

b. peak voltage   2. the maximum voltage reached by the signal

c. peak-peak voltage   3. twice the peak voltage (amplitude)

d. frequency   4. a voltage or current which conveys information

Chinese Translation of Texts（参考译文）

第3课　交流电、直流电及电信号

交流电

交流电流的方向会从一个方向转换成另一个方向，并且持续不断地转换（见图3-1和图3-2）。交流电压会持续不断地在正负两个极性之间交替变化。这种方向变化的速率称为交流信号的频率，单位是赫兹，它表示一秒内周期性变化的次数。

交流电源适用于给一些设备供电，如灯和加热器，但几乎所有的电子电路都需要稳定的直流电源。

直流电

直流电流总是向同一个方向流动，但它可能增大或者减小。直流电压总是正的（或者总是负的），但它也可能增大或者减小（见图3-3）。电子电路通常需要一个具有定值的稳定的直流电源。

干电池、蓄电池（组）和稳压电源能提供对于电子电路来说理想的、稳定的直流电源。

在直流电源的作用下，电灯、加热器和发动机可以正常工作。

电信号的特性

电信号是传送信息的电压或者电流，通常是指电压。这个术语可以被用于电路里的任何电压或者电流。

图3-4的电压-时间图显示了电信号的各种特性。除了图上标明的那些特性以外还有频率，它表示每秒钟的周期数。该图显示了一个正弦波，但这些特性适用于具有恒定形状的任何信号。

振幅是信号所达到的最大电压。单位是伏特，V。峰值电压是振幅的另一个名称。峰-峰值电压是峰值电压（振幅）的两倍。当对示波器上的轨迹进行读数时，我们通常测量峰-峰值电压。

周期是信号完成一个周期所需要的时间。周期的单位是秒，但是周期往往非常短，所以常用毫秒（ms）和微秒（μs）来表示。1 ms = 0.001 s，1 μs = 0.000001 s。

频率是每秒的周期数。单位是赫兹（Hz），但频率往往比较高，所以常用千赫（kHz）和兆赫（MHz）来表示。频率＝1/周期，周期＝1/频率。

另一个常用的值是有效值，电压或电流的有效值作用于电阻时，与相同大小的直流电压或电流作用于相同大小的电阻等效。

Lesson 4 Transistor Voltage Amplifier

Amplifiers are necessary in many types of electronic equipment such as radios, oscilloscopes and record players. Often it is a small alternating voltage that has to be amplified. A junction transistor in the common-emitter mode can act as a voltage amplifier if a suitable resistor, called the load, is connected in the collector circuit.

The small alternating voltage, the input u_i, is applied to the base-emitter circuit and causes small changes of base current which produce large changes in the collector current flowing through the load. The load converts these current changes into voltage changes which form the alternating output voltage u_o (u_o being much greater than u_i).

A transistor voltage amplifier circuit is shown in Fig.4-1. To see just the voltage amplification occurs, consider first the situation when there is no input, i.e. $u_i = 0$, called the quiescent (quiet) state.

For transistor action to take place the base-emitter junction must always be forward biased. A simple way of ensuring this is to connect a resistor R_B, called the base-bias resistor. A steady (DC) base current I_B flows from battery "+", through R_B into the base and back to battery "−", via the emitter. The value of R_B can be calculated once the value of I_B for the best amplifier performance has been decided.

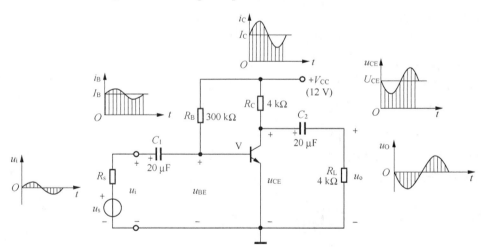

Fig.4-1 Transistor Voltage Amplifier

If V_{CC} is the battery voltage and V_{BE} is the base-emitter junction voltage (always about +0.6V for an n-p-n silicon transistor), then for the base-emitter circuit, since DC voltages add up, we can write

$$V_{CC} = I_B \times R_B + V_{BE} \tag{1}$$

I_B causes a much larger collector current I_C which produces a voltage drop $I_C \times R_L$ across the load R_L. If V_{CE} is the collector-emitter voltage, then for the collector-emitter circuit

$$V_{CC} = I_C \times R_L + V_{CE} \tag{2}$$

When u_i is applied and goes positive, it increases the base-emitter voltage slightly (e.g. from +0.60V to +0.61V). When u_i swings negative, the base-emitter voltage decreases slightly (e.g. from +0.60V to +0.59V). As a result a small alternating current is superimposed on the quiescent base current I_B which in effect becomes a varying DC.

When the base current increases, large proportionate increases occur in the collector current. From equation (2) it follows that there is a corresponding large decrease in the collector-emitter voltage (since V_{CC} is fixed). A decrease of base current causes a large increase of collector-emitter voltage. In practice positive and negative swings of a few milli-volts in u_i, can result in a fall or rise of several volts in the voltage across R_L and so also in the collector-emitter voltage.

The collector-emitter voltage may be regarded as an alternating voltage superimposed on a steady direct voltage, i.e. on the quiescent value of V_{CE}. Only the alternating part is wanted and capacitor C blocks the direct part but allows the alternating part, i. e. the output u_o to pass.

New Words

oscilloscope [ə'siləskəup] n.	示波器
alternating ['ɔ:ltəneitiŋ] adj.	交互的
amplifier ['æmplifaiə] n.	放大器
junction ['dʒʌŋkʃn] n.	连接，接合，交叉点，汇合处
resistor [ri'zistə] n.	电阻器
quiescent [kwi'esnt] adj.	静止的
biased ['baiəst] adj.	结果偏倚的，有偏的
silicon ['silikən] n.	硅，硅元素
superimpose [,su:pərim'pəuz] v.	添加，双重
proportionate [prə'pɔ:ʃənət] adj.	成比例的
equation [i'kweiʃn] n.	方程式，等式

Phrases and Expressions

to convert…into	把……转变成
as a result	结果

Notes

1. Often it is a small alternating voltage that has to be amplified.
   译文：通常小的交流电压需要放大。
   这是一个常见句型，it 是形式主语，that has to be amplified 引导真正意义上的主语从句。

2. …which produce large changes in the collector current flowing through the load.

(1) which 引出 base current。(2) flowing through the load 是现在分词短语，充当后置定语，修饰 collector current。

3. The load converts these current changes into voltage changes which form the alternating output voltage u_o (u_o being much greater than u_i).

译文：负载将交变电流转变成交变电压，即输出交流电压 u_o（u_o 比 u_i 大）。

(1) to convert … into … 把……变成……，相当于 to change … into… (2) which form the alternating output voltage u_o 是定语从句，修饰 voltage changes。(3) alternating 是现在分词，修饰 output voltage。

4. A simple way of ensuring this is to connect a resistor R_B, called the base-bias resistor.

译文：简单的方法是接一个电阻 R_B，称为基极偏置电阻。

不定式 to connect a resistor R_B 充当表语，其中 called the base-bias resistor 是过去分词短语，充当定语。

5. As a result a small alternating current is superimposed on the quiescent base current I_B which in effect becomes a varying DC.

译文：结果是一个小的交流电流叠加在静态基极电流 I_B 上并产生一个变化的直流电压。

(1) as a result: "结果"。(2) which in effect becomes a varying DC 是定语从句，修饰 the quiescent base current I_B。

6. … and so also in the collector-emitter voltage.

so 这里指 can also result in a fall or rise of several volts。

7. The collector-emitter voltage may be regarded as an alternating voltage superimposed on a steady direct voltage, i.e. on the quiescent value of V_{CE}.

译文：集-射间电压可以认为是一个交流电压叠加在静态直流电压（即静态电压 V_{CE}）之上。

过去分词短语 superimposed on a steady direct voltage, i.e. on the quiescent value of V_{CE} 充当定语，修饰 an alternating voltage。

Exercises

1. Write T (True) or F (False) beside the following statements about the text.

a. Amplifiers are necessary in many types of electronic equipment such as radios, oscilloscopes and record players. Often it is a large alternating current that has to be amplified.

b. A junction transistor in the common-emitter mode can act as a resistance amplifier.

c. The small alternating voltage is applied to the base-emitter circuit and causes large changes of base current which produce small changes in the collector current flowing through the load.

d. The load converts current changes into voltage changes, which form the alternating output voltage.

e. For transistor action to take place the base-emitter junction must always be forward straight line.

f. When the base current increases, large proportionate increases occur in the collector current.

g. A decrease of base current causes a large increase of collector-emitter voltage.

h. The collector-emitter voltage may be regarded as an alternating voltage superimposed on a steady direct voltage.

2. Match the following terms to appropriate definition or expression.

a. load   1. a suitable resistor
b. V_{CC} 2. the alternating output voltage
c. u_{BE} 3. base-emitter junction voltage
d. u_o 4. the base-bias resistor
e. u_i 5. battery voltage
f. R_B 6. the small alternating voltage, the input

3. Fill in the missing words according to the text.

a. A junction transistor in the common-emitter mode can act as a _____ if a suitable resistor called the _____, is connected in the collector circuit.

b. The load converts these _____ changes into _____ changes, which form the alternating output voltage u_o.

c. A simple way of ensuring this is to connect a resistor R_B, called the _____. A steady (DC) base current I_B flows from battery "+", through R_B into the base and back to battery "–", via the _____.

d. The collector-emitter voltage may be regarded as an _____ superimposed on a steady direct voltage, i.e. on the quiescent value of V_{CE}. Only the alternating part is wanted and capacitor C blocks the direct part but allows the _____, i.e. the output u_o to pass.

4. Translate the following paragraph into Chinese.

When the base current increases, large proportionate increases occur in the collector current. From equation (2) it follows that there is a corresponding large decrease in the collector-emitter voltage (since V_{CC} is fixed). A decrease of base current causes a large increase of collector-emitter voltage. In practice positive and negative swings of a few milli-volts in u_i, can result in a fall or rise of several volts in the voltage across R_L and so also in the collector-emitter voltage.

Chinese Translation of Texts（参考译文）

第 4 课　晶体管电压放大器

通常小的交流电压需要放大，所以放大器广泛用于各种电子设备中，如收音机、示波器和录音机等。一个结型晶体管连接成共发射极组态，如果在其集电极回路中接入合适的电阻，这个电阻被称为负载，则可构成电压放大器。

一个小的交流输入电压 u_i，加在基-射回路中，可产生小的基极交变电流，进而产生一个大的集电极交变电流流过负载，负载将交变电流转变成交变电压，即输出交流电压 u_o（u_o 比 u_i 大）。

一个简单的电路如图 4-1 所示,表明了电压放大的过程,首先考虑没有输入,即 $u_i=0$ 的情形,称为静态。

晶体管工作在放大状态时其发射结必须为正向偏置。为保证发射结正偏,简单的方法是接一个电阻 R_B,称为基极偏置电阻。基极偏置电流 I_B 从电源正极流出,经 R_B 流入基极;再经发射极流回电源负极。一旦 I_B 在放大器的最佳状态下确定后,可以随之计算出 R_B 的值。

如果 V_{CC} 为电源电压,V_{BE} 为发射结电压(n-p-n 硅管为 0.6V),则对基极回路,若加有直流电压,有

$$V_{CC} = I_B \times R_B + V_{BE} \tag{1}$$

I_B 产生了一个相当大的集电极电流 I_C,其在负载 R_L 上产生电压降 $I_C \times R_L$。若 V_{CE} 为集电结电压,则对集电极回路,有

$$V_{CC} = I_C \times R_L + V_{CE} \tag{2}$$

当交流电压 u_i 加入且为正时,基极电压略有增加(如从 0.60V 到 0.61V);当 u_i 为负时,基极电压略有下降(如从 0.60V 到 0.59V)。结果是一个小的交流电流叠加在静态基极电流 I_B 上并产生一个变化的直流电压。

基极电流增大时,集电极电流会大幅增大。由式(2)可知,将相应引起集-射间电压大幅下降(因为 V_{CC} 一定)。同理,基极电流减小将引起集-射间电压大幅增加,即 u_i 在正负间有毫伏数量级的摆动,就会导致负载 R_L 上的电压及集-射间电压有较大的幅值变化。

集-射间电压可以认为是一个交流电压叠加在静态直流电压(即静态电压 V_{CE})之上,若只需交流部分,则可利用电容器 C 的隔直作用,将直流部分去掉,只允许交流电压 u_o 通过。

Lesson 5 Digital Circuit

The phrase "digital electronics" is used to describe those circuit systems which primarily operate with the use of only two different voltage levels or two other binary states. The two different states by which digital circuits operate may be of several forms. They can, in the simplest form, consist of the opening and closing of a switch. In this case, the closed-switch state can be represented by 1 and the open-switch state by 0.

A very common method of digital operation is achieved by using voltage pulses. The presence of a positive pulse can be represented by 1 and the absence of a pulse by 0. With a square-wave signal, the positive pulses can represent 1 and the negative pulses can represent 0.

Integrated circuits containing many transistors are most commonly used as switching devices in digital electronics logic gates. The three basic types of digital logic gates are the AND gate, the OR gate, and the NOT gate (as shown in Fig.5-1). The operation of an AND gate is mathematically expressed by the equation $A \times B = C$. This can be read as "input A and input B equals output C". The operation of an OR gate is often expressed by the equation $A + B = C$. This can be read as "either input A or input B (or both) equals output C". An important function of NOT gate is to produce signal inversion or an output signal that is opposite in nature to the input signal. Any logic function can be performed by the three basic gates that have been described. Even in a large scale digital system, such as a computer, control, or digital-communication system, there are only a few basic operations, which must be performed. The three basic types of digital logic gates and the flip-flop are the four circuits most commonly employed in such systems.

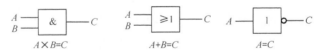

Fig.5-1 Digital Logic Gates

Another form of digital circuit is programmable logic devices. Programmable logic devices can perform the same functions as machines based on logic gates, but can be easily reprogrammed without changing the wiring. This means that a designer can often repair design errors without changing the arrangement of wires. Therefore, in small volume products, programmable logic devices are often the preferred solution. They are usually designed by engineers using electronic design automation software.

New Words

primarily ['praimərəli] adv.　　　　首先，起初，主要地，根本上
binary ['bainəri] adj.　　　　　　　二进制的，二元的
switch [switʃ] n. & vt.　　　　　　 开关；转换
achieve [ə'tʃi:v] vt.　　　　　　　　完成，达到

pulse [pʌls] n.	脉搏，脉冲
function ['fʌŋkʃən] n. & vi.	功能，函数；运行
inversion [in'və:ʃən] n.	倒置
scale [skeil] n.	刻度，衡量，比例，数值范围
flip-flop ['flipflɔp] n.	触发器

Phrases and Expressions

integrated circuits	集成电路

Notes

1. The phrase "digital electronics" is used to describe those circuit systems which primarily operate with the use of only two different voltage levels or two other binary states.

   译文："数字电子技术"这一术语是用来描述只用两种不同的电压、电平或两种不同的二进制状态进行工作的电路系统的。

   (1) to describe … or two other binary states 是不定式短语，充当目的状语。(2) which primarily operate with the use of only two different voltage levels or two other binary states 是定语从句，修饰 systems。

2. The two different states by which digital circuits operate may be of several forms.

   译文：数字电路用以工作的两种不同状态有几种形式。

   句中 by which digital circuits operate 是定语从句，修饰 states。

3. In this case, the closed-switch state can be represented by 1 and the open-switch state by 0.

   译文：在这种情况下，开关闭合状态用 1 表示，开关断开状态用 0 表示。

   (1) closed-switch 和 open-switch 具有形容词性，充当定语，均修饰 state。(2) and the open-switch state by 0 已省略了谓语动词部分 can be represented。

4. with a square-wave signal, … 若用方波信号，……

   这是一个表示条件的介词短语，其作用相当于一个条件状语从句。

5. An important function of NOT gate is to produce signal inversion or an output signal that is opposite in nature to the input signal.

   译文：非门的重要功能是产生反向信号，即产生与输入信号性质相反的输出信号。

   不定式 to produce signal inversion or an output signal 充当表语，其后是定语从句 that is opposite in nature to the input signal，用以修饰该表语。

6. …such as a computer, control, or digital-communication system, there are only a few basic operations, which must be performed.

   译文：……例如，计算机系统、控制系统或数字通信系统中，需要进行的基本运算也只有这几种。

   (1) such as 引出 a large scale digital system 的同位语。(2) which must be performed 是非限制

性定语从句。

7. The three basic types of digital logic gates and the flip-flop are the four circuits most commonly employed in such systems.

译文：这3种基本形式的数字逻辑门和触发器是这些系统中最常用的4个电路。

过去分词短语 most commonly employed in such systems 充当定语，修饰 circuits，其中 employed 在这里的意义是"使用"，相当于 used。

Exercises

1. Write T (True) or F (False) beside the following statements about the text.

a. The two different states by which digital circuits operate is only one form.

b. The closed-switch state can be represented by 1 and the open-switch state can be represented by 2.

c. A very common method of digital operation is achieved by using current pulses.

d. Integrated circuits containing many transistors are most commonly used as switching devices in digital electronics logic gates.

e. The three basic types of digital logic gates are the AND gate, the OR gate, and the NOT gate.

f. An important function of NOT gate is to produce signal inversion or an input signal that is opposite in nature to the output signal.

g. The three basic types of digital logic gates and the flip-flop are the four circuits most commonly employed in such systems.

h. Even in a large scale digital system, such as a computer, control, or digital-communication system, there are a lot of basic operations, which must be performed.

2. Match the following terms to appropriate definition or expression.

a. digital electronics　　　　1. circuit systems which primarily operate with the use of only two different voltage levels or two other binary states

b. $A \times B = C$　　　　2. input A and input B equals output C

c. $A + B = C$　　　　3. either input A or input B (or both) equals output C

d. voltage pulses　　　　4. a very common method of digital operation

3. Fill in the missing words according to the text.

a. They can, in the simplest form, consist of the opening and closing of a switch. In this case, _____ can be represented by 1 and _____ by 0.

b. The presence of a _____ also can be represented by 1 and the _____ of a pulse by 0.

c. With a _____ signal, the positive pulses can represent 1, the negative pulses can represent 0.

d. The operation of an OR gate is often expressed by the equation _____.

4. Translate the following paragraphs into Chinese.

Digital signals and circuits are the vast and important subject. Digital signals are binary in nature taking on values in one of two well-defined ranges. The set of basic operations that can be performed on digital signals is quite small. The behavior of any digital system, (up to and including the most sophisticated digital computer) can be represented by appropriate combinations of digital variables and the digital operations from this small set.

We concerned with digital system variables that take on only two values (binary variables). We conventionally denote these values as "0" or "1", and then use a special set of rules called Boolean algebra to summarize the various ways in which digital variables can be combined. This algebra and much of the notation are adopted directly from mathematical logic. Thus, "logic variable" or "logic operation" are commonly used in place of digital variable, or digital operation.

Chinese Translation of Texts（参考译文）

第 5 课　数字电路

"数字电子技术"这一术语是用来描述只用两种不同的电压、电平或两种不同的二进制状态进行工作的电路系统的。数字电路用以工作的两种不同状态有几种形式。最简单的一种形式是由开关的断开和闭合状态组成的。在这种情况下，开关闭合状态用 1 表示，开关断开状态用 0 表示。

最常用的数字运算法是用电压脉冲完成的。有正脉冲存在用 1 表示，没有正脉冲存在用 0 表示。若用方波信号，则正脉冲可代表 1，负脉冲可代表 0。

含有许多晶体管的集成电路被广泛用作电子数字逻辑门中的开关器件。有 3 种基本逻辑门，它们是与门、或门和非门（见图 5-1）。与门运算用数学等式可表示为 $A \times B = C$，读作"输入 A 与输入 B 等于输出 C"。或门运算常用方程 $A+B=C$ 表示，读作"输入 A 或输入 B（或两者）等于输出 C"。非门的重要功能是产生反向信号，即产生与输入信号性质相反的输出信号。任何一种逻辑功能都可用上述 3 种基本门电路来完成。即使是在大规模数字系统，例如，计算机系统、控制系统或数字通信系统中，需要进行的基本运算也只有这几种。3 种基本形式的数字逻辑门和触发器是这些系统中最常用的 4 种电路。

另一种形式的数字电路是可编程逻辑器件，可编程逻辑器件可以完成与基于逻辑门的电路相同的功能，但可以很容易地重新编程而不改变布线。这意味着设计师往往可以修复设计错误而不改变布线。因此，在体积较小的产品中，可编程逻辑器件往往是首选的解决方案。它们通常是由设计工程师利用电子设计自动化软件设计完成的。

Lesson 6　Circuit Diagrams and Block Diagrams

　　Circuit diagrams show the connections as clearly as possible with all wires drawn neatly as straight lines. The actual layout of the components is usually quite different from the circuit diagram, and this can be confusing for the beginner. The secret is to concentrate on the connections, not the actual positions of components.

　　A circuit diagram is useful when testing a circuit and for understanding how it works. This is why the instructions for projects include a circuit diagram as well as the printed circuit board layout which you need to build the circuit.

　　Drawing circuit diagrams is not difficult but it takes a little practice to draw neat, clear diagrams (as shown in Fig.6-1). This is a useful skill for science as well as for electronics. You will certainly need to draw circuit diagrams if you design your own circuits.

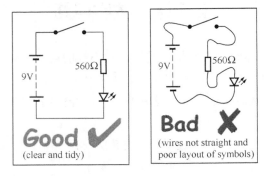

Fig.6-1　Drawing Circuit Diagrams

Follow these tips for best results.
- Make sure you use the correct symbol for each component.
- Draw connecting wires as straight lines (use a ruler).
- Label components such as resistors and capacitors with their values.
- The positive (+) supply should be at the top and the negative (−) supply at the bottom.

　　Block diagrams are used to understand (and design) complete circuits by breaking them down into smaller sections or blocks. Each block performs a particular function, and the block diagram shows how they are connected together. No attempt is made to show the components used within a block, and only the inputs and outputs are shown. This way of looking at circuits is called the systems approach.

　　Power supply (or battery) connections are usually not shown on block diagrams.

　　Audio Amplifier System (as shown in Fig.6-2)

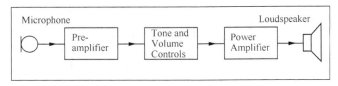

Fig.6-2　Block Diagram of an Audio Amplifier System

- **Microphone**—a transducer which converts sound to voltage.
- **Pre-amplifier**—amplifies the small audio signal (voltage) from the microphone.
- **Tone and Volume Controls**—adjust the nature of the audio signal.

The tone control adjusts the balance of high and low frequencies.

The volume control adjusts the strength of the signal.

- **Power Amplifier**—increases the strength (power) of the audio signal.
- **Loudspeaker**—a transducer which converts the audio signal to sound.

Radio Receiver System (as shown in Fig.6-3)

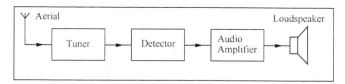

Fig.6-3 Block Diagram of a Radio Receiver System

- **Aerial**—picks up radio signals from many stations.
- **Tuner**—selects the signal from just one radio station.
- **Detector**—extracts the audio signal carried by the radio signal.
- **Audio Amplifier**—increases the strength (power) of the audio signal.

This could be broken down into the blocks like the Audio Amplifier System shown above.

- **Loudspeaker**—a transducer which converts the audio signal to sound.

Feedback Control System (as shown in Fig.6-4)

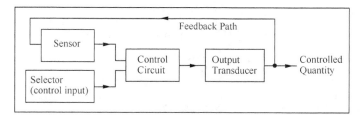

Fig.6-4 Block Diagram of a Feedback Control System

- **Sensor**—a transducer which converts the state of the controlled quantity to an electrical signal.
- **Selector (control input)**—selects the desired state of the output. Usually it is a variable resistor.
- **Control Circuit**—compares the desired state (control input) with the actual state (sensor) of the controlled quantity and sends an appropriate signal to the output transducer.
- **Output Transducer**—converts the electrical signal to the controlled quantity.
- **Controlled Quantity**—usually not an electrical quantity, e.g. motor speed.
- **Feedback Path**—usually not electrical, the sensor detects the state of the controlled quantity.

New Words

diagram ['daiəgræm] n.	图表，图解
layout ['leiaut] n.	布局，安排，布置图，规划图
instruction [in'strʌkʃn] n.	指令，说明
neat [ni:t] adj.	整洁的
straight [streit] adj.	直的
aerial ['eəriəl] n.	天线
tuner ['tju:nə] n.	调谐器
detector [di'tektə] n.	检波器
loudspeaker [ˌlaud'spi:kə] n.	扬声器，扩声器，喇叭
sensor ['sensə] n.	传感器

Phrases and Expressions

block diagram	方框图
break down	分解
pre-amplifier	前置放大器
audio amplifier	音频放大器
volume control	音量控制
power amplifier	功率放大器
feedback path	反馈路径
control circuit	控制电路

Notes

1. This is why the instructions for projects include a circuit diagram as well as the printed circuit board layout which you need to build the circuit.

   译文：这就是项目说明既包括电路图又包括建立实际电路所需要的印刷电路板的布局图的原因。

2. Block diagrams are used to understand (and design) complete circuits by breaking them down into smaller sections or blocks.

   译文：方框图是通过把完整的电路分解成较小的模块来理解（设计）电路的。

   break down into: 分解成……

Unit I Basic Knowledge of Electronics 电子技术基础知识

Exercises

1. Match the following terms to appropriate definition or expression.

a. detector   1. a transducer which converts sound to voltage

b. microphone   2. extracts the audio signal carried by the radio signal

c. tone and volume controls   3. increases the strength (power) of the audio signal

d. power amplifier   4. adjust the nature of the audio signal

e. tuner   5. selects the signal from just one radio station

2. Fill in the missing words according to the text.

a. Circuit diagrams show the connections as _____ as possible with all wires _____ as straight lines.

b. A circuit diagram is useful when _____ and for understanding how it works.

c. Label components such as _____ and capacitors with their values.

d. Control Circuit—compares _____ (control input) with the _____ of the controlled quantity and sends an appropriate signal to the output transducer.

e. Drawing circuit diagrams is not difficult but it takes a little practice to _____, _____ diagrams.

3. Translate the following paragraph into Chinese.

Block diagrams are used to understand (and design) complete circuits by breaking them down into smaller sections or blocks. Each block performs a particular function, and the block diagram shows how they are connected together. No attempt is made to show the components used within a block, and only the inputs and outputs are shown. This way of looking at circuits is called the systems approach.

Chinese Translation of Texts（参考译文）

第6课 电路图和方框图

电路图要尽可能清晰地显示电路的连接，所有的连接线都要画成整齐的直线。实际的元件布局通常与电路图不同，这可能使初学者感到迷惑。其奥秘在于连接线，而不是元件的实际位置。

电路图在测试电路和理解它是如何工作的时候是有用的。这就是项目说明既包括电路图又包括建立实际电路所需要的印刷电路板的布局图的原因。

绘制电路图

绘制电路图并不难，但绘制工整、清晰的图表需要一些练习（见图6-1）。这对科学及电子学而言都是一个有用的技能。如果自己设计电路，肯定需要绘制电路图。

按照这些小建议将会取得最好的效果。
- 确保对每个元件都使用了正确的符号。
- 将连接线绘制成直线（请使用直尺）。
- 标注元件的值，如电阻和电容的值。
- 正电源应放在顶部，而负电源应放在底部。

方框图是通过把完整的电路分解成较小的模块来理解（设计）电路的。每个模块执行某一特定功能，方框图显示了它们是如何连接在一起的。不需要画出每一个模块所采用的具体元件，只需画出其输入和输出。这种对待电路的方式被称为系统方法。

电源（或电池）的连接通常不会在方框图中画出。

音频放大器系统（见图6-2）
- 麦克风：将声音转换成电压的换能器。
- 前置放大器：放大来自于麦克风的小的音频信号（电压）。
- 音调和音量控制调节：调节音频信号的参数。

音调控制调节平衡频率的高低。
音量控制调节信号的强度。
- 功率放大器：增加音频信号的强度（功率）。
- 扬声器：将音频信号转换为声音的换能器。

无线电接收系统（见图6-3）
- 天线：从许多无线电台中收集无线电信号。
- 调谐器：选出一个无线电台的信号。
- 检波器：将音频信号从携带它的无线电信号中提取出来。
- 音频放大器：增加音频信号的强度（功率）。

这可以被分解成很多块，就像上面的音频放大系统显示的。
- 扬声器：将音频信号转换为声音的换能器。

反馈控制系统（见图6-4）
- 传感器：将控制量转换成电子信号的换能器。
- 选择器（控制输入）：选择所需的输出状态。通常它是一个可变电阻。
- 控制电路：将理想状态（控制输入）与被控制量的实际状态（传感器）进行比较，然后发送一个合适的信号给输出换能器。
- 输出换能器：将电信号转换成控制量。
- 控制量：通常不是一个电气量，如电机速度。
- 反馈路径：传感器检测被控量的状态，通常不是电气量。

Reading Material

1. Basic Resistor Circuits

Circuits consisting of just one battery and one load resistance are very simple to analyze, but they are not often found in practical applications. Usually, we find circuits where more than two components are connected together. There are two basic ways in which resistors can be connected: in series and in parallel. A simple series resistance circuit is shown in Fig.R1-1.

Determining the total resistance for two or more resistors in series is very simple. Total resistance equals the sum of the individual resistances. In this case, $R_T=R_1+R_2$. This makes common sense, and if you think again in terms of water flow, a series of obstructions in a pipe add up to slow the flow more than any one. The resistance of a series combination is always greater than any of the individual resistors.

The other method of connecting resistors is shown in Fig.R1-2, which shows a simple parallel resistance circuit.

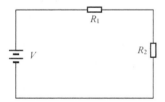
Fig.R1-1   Two Resistors in Series

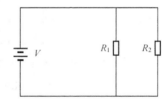
Fig.R1-2   Two Resistors in Parallel

Our water pipe analogy indicates that it should be easier for current to flow through this multiplicity of paths, even easier than it would be to flow through any single path. Thus, we expect a parallel combination of resistors to have less resistance than any one of the resistors. Some of the total current will flow through R_1, and some will flow through R_2, causing an equal voltage drop across each resistor. More current, however, will flow through the path of least resistance. The formula for total resistance in a parallel circuit is more complex than for a series circuit.

$$\frac{1}{R_T}=\frac{1}{R_1}+\frac{1}{R_2}$$

Parallel and series circuits can be combined to make more complex structures, but the resulting complex resistor circuits can be broken down and analyzed in terms of simple series or parallel circuits. Why would you want to use such combinations? There are several reasons. You might use a combination to get a value of resistance that you needed but did not have in a single resistor. Resistors have a maximum voltage rating, so a series of resistors might be used across a high voltage. Also, several low power resistors can be combined to handle higher power. What type of connection would you use?

And, of course, the complexity doesn't stop at simple series and parallel either! We can have circuits that are a combination of series and parallel, too.

In this circuit (as shown in Fig.R1-3), we have two loops for electrons to flow through: one from 6 to 5 to 2 to 1 and back to 6 again, and another from 6 to 5 to 4 to 3 to 2 to 1 and back to 6 again. Notice how both current paths go through R_1 (from point 2 to point 1). In this configuration, we'd say that R_2 and R_3 are in parallel with each other, while R_1 is in series with the parallel combination of R_2 and R_3.

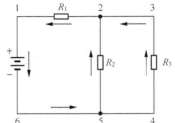

Fig.R1-3　Series-parallel Circuit

2. Capacitance and Inductance

Capacitance

Electrical energy can be stored in an electric field. The device capable of doing this is called a capacitor or a condenser.

A simple condenser consists of two metallic plates separated by a dielectric. If a condenser is connected to a battery, the electrons will flow out of the negative terminal of the battery and accumulate on the condenser plate connected to that side. At the same time, the electrons will leave the plate connected to the positive terminal and flow into the battery to make the potential difference just the same as that of the battery (as shown in Fig.R2-1 (a)). Thus the condenser is said to be charged.

The capacitance is directly proportional to the dielectric constant of the material and to the area of the plates and inversely to the distance of the plates. It is measured in farads. When a change of one volt per second across it causes the current of one ampere to flow, the condenser is said to have the capacitance of one farad. However, farad is too large for a unit to be used in radio calculation, so microfarad (one millionth of a farad) and the micromicrofarad (10^{-12} farads) are generally used.

The amount of the stored energy of a charged condenser is proportional to the applied voltage and its capacitance. The capacitance of a condenser is determined by three important factors, namely, the area of the plate surface, the space between them and dielectric material. The larger the plate area, the smaller the space between them, the greater the capacitance.

One of the condensers to be used in radio receiver is a variable condenser, whose capacitance can be varied by turning the plates. It is used in the receiver for tuning and varying capacitance in the circuit so as to pick up the desired signals of different wavelengths.

Inductance

It is well known that inductors are one of the main building blocks in electronic circuits. An inductor is simply a coil of wire with or without a magnetic core (as shown in Fig.R2-1(b)).

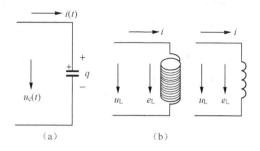

Fig.R2-1   Capacitance and Inductance

All coils have inductance. Inductance is the property of opposing any change of current flowing through a coil. If a coil offers a large opposition to the current flowing through it at a certain frequency, it is said to have large inductance. A small inductance would provide less opposition at the same frequency, and resistance offers an opposition to all current flow.

When an emf is applied to a coil, there is an induced emf in it. The polarity of the induced emf is always such as to oppose any change in the current in the circuit. This means that the property of inductance oppose an increase in current just as much as it opposes a decrease in current.

A coil of many turns will have more inductance than one of few turns. Also if a coil is placed on an iron core its inductance will be greater than it was without the magnetic core. The unit of inductance is the henry. A coil has an inductance of one henry if an induced emf of one volt is induced in the coil when the current through it changes at the rate of one ampere per second. Values of inductance used in radio equipment vary over a wide range.

3. Diodes and Its Circuit

Both p-type and n-type silicon will conduct electricity just like any conductor; however, if a piece of silicon is doped p-type in one section and n-type in an adjacent section, current will flow in only one direction across the junction between the two regions. This device is called a diode and is one of the most basic semiconductor devices.

A diode is called forward biased if it has a positive voltage across it from the p-type to n-type material. In this condition, the diode acts rather like a good conductor, and current can flow, as in Fig.R3-1.

There will be a small voltage across the diode, about 0.6 volts for Si, and this voltage will be largely independent of the current, very different from a resistor.

If the polarity of the applied voltage is reversed, then the diode will be reversing biased and will appear nonconducting (Fig.R3-2). Almost no current will flow and there will be a large voltage across the device.

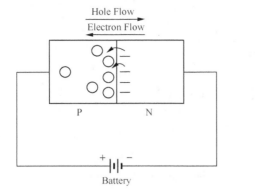
Fig.R3-1 A Forward Biased Diode

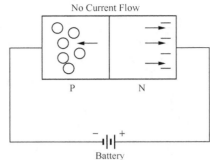
Fig.R3-2 A Reverse Biased Diode

The diode acts like a one-way valve for current, and this is a very useful characteristic. One application is to convert alternating current (AC), which changes polarity periodically, into direct current (DC), which always has the same polarity. Normal household power is AC while batteries provide DC, and converting from AC to DC is called rectification. Diodes are used so commonly for this purpose that they are sometimes called rectifiers, and although there are other types of rectifying devices. Fig.R3-3 shows the input and output current for a simple half-wave rectifier. The circuits gets its name from the fact that the output is just the positive half of the input waveform. A full-wave rectifier circuit (shown in Fig.R3-4) uses four diodes arranged so that both polarities of the input waveform can be used at the output.

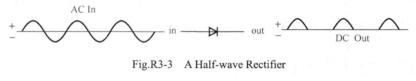
Fig.R3-3 A Half-wave Rectifier

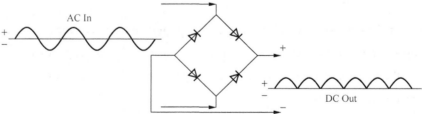
Fig.R3-4 A Full-wave Rectifier

4. The Transistor and Its Basic Circuit

Transistors are the most important device in electronics today. Not only are they made as discrete (separate) components but integrated circuits (ICs) may contain several thousand on a tiny slice of silicon.

Transistors are three-terminal devices used as amplifiers and as switches. There are two basic types.

(Ⅰ) the junction transistor (usually called the transistor); its operation depends on the flow of both majority and minority carriers, and it has two PN junctions.

(Ⅱ) the field effect transistor (called the FET) in which current is due to majority carriers only (either electrons or holes), and there is just one PN junction.

A transistor consists of three layers of semiconductor material: a thin layer of one type with the other type on each side. There are two possible arrangements: N-type in the middle with P-type on each side (PNP) and P-type in the middle with N-type on each side (NPN). The center is called base, one outside layer is called the emitter, and the other is known as the collector (as shown in Fig.R4-1).

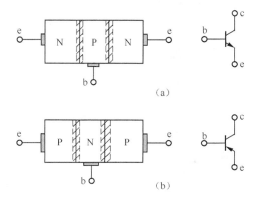

Fig.R4-1　PN Junction and Transistor

A transistor is an electronically device that regulates the current flowing through it. Current from a power source enters the emitter, passes through the very thin base region, and leaves via the collector. Current flow is always in this direction. This current can be made to vary in amplitude by varying the current flowing in the base circuit. It takes only a small change of base current to control a relatively large collector current. It is this ability that enables the transistor to amplify.

There are three basic ways of connecting transistors in a circuit (as shown in Fig.R4-2): common-base, common-emitter, and common-collector. In the common-base connection, the signal is introduced into the emitter-base circuit and extracted from the collector-base circuit. Because the input or emitter-base circuit has a low impedance in the order of 0.5 to 50 ohms, and the output or collector base circuit has a high impedance in the order of 1000 ohms to one megohm, the voltage or power gain in this type of configuration may be in the order of 1500.

In the common-emitter connection, the signal is introduced into the base-emitter circuit and extracted from the collector-emitter circuit. This configuration has more moderate input and output impedance than the common-base circuit. The input (base-emitter) impedance is in the range of 20 to 5,000 ohms, and output (collector-emitter) impedance is about 50 to 50,000 ohms. Power gains in the order of 10,000 (or about 40 db) can be realized with this circuit because it provides both current gain and voltage gain.

The third type of connection is the common-collector circuit. In this configuration, the signal is introduced into the base-collector circuit and extracted from the emitter-collector circuit. Because the input impedance of the transistor is high and the output impedance low in this connection, the voltage gain is less than one and the power gain is usually lower than that obtained in a common-base or a common-emitter circuit.

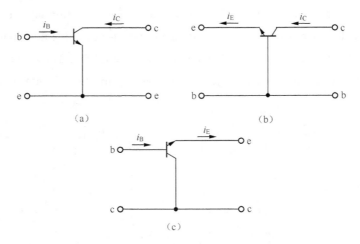

Fig.R4-2 Three Basic Ways of Connecting Transistors in a Circuit

5. Radio Waves

Radio Waves are a member of the electromagnetic family of waves. They are energy-carriers which travel at the speed of light (v), their frequency (f) and wavelength (λ) being related, as for any wave motion, by the equation

$$v = f \times \lambda$$

where $v = c = 3.0 \times 10^8$ m/s in a vacuum (or air). If $\lambda = 300$ m, then $f = v/\lambda = 3.0 \times 10^8/(3.0 \times 10^2) = 10^6$ Hz = 1 MHz. The smaller λ is, the larger f.

Radio Waves can be described either by their frequency or their wavelength, but the former is more fundamental since, unlike λ (and v), f does not change when the waves travel from one medium to another.

Radio Waves can travel from a transmitting aerial in one or more of three different ways (as shown in Fig.R5-1).

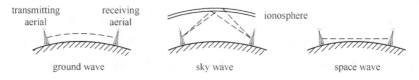

Fig.R5-1 Three Different Radio Waves

(a) Surface or ground wave. This travels along a ground, following the curvature of the Earth's surface. Its range is limited mainly by the extent to which energy is absorbed from it by the ground. Poor conductors such as sand absorb more strongly than water, and the higher the frequency the greater the absorption. The range may be about 1,500 km at low frequencies (long wave, but much less for v.h.f).

(b) Sky wave. This travels skywards and, if it is below a certain critical frequency (typically 30 MHz), is returned to earth by the ionosphere. This consists of layers of air molecules (the D, E and F layers), stretching from about 80 km above the Earth to 500 km, which have become positively charged through the removal of electrons by the sun's ultraviolet radiation. On striking the Earth the sky wave bounces back to the ionosphere where it is again gradually refracted and returned earthwards as if by 'reflection'. This continues until it is completely attenuated.

(c) Space wave. For v. h. f., u.h.f. and microwave signals, only the space wave, giving line-of-sight transmission, is effective. A range of up to 150 km is possible on Earth if the transmitting aerial is on high ground and there are no intervening obstacles such as hills, buildings or trees.

Radio Waves have frequencies from 300 GHz to as low as 3 Hz, and corresponding wavelengths from 1 millimeter to 100 kilometers. Like all other electromagnetic waves, they travel at the speed of light. Naturally occurring radio waves are made by lightning, or by astronomical objects. Artificially generated radio waves are used for fixed and mobile radio communication, broadcasting, radar and other navigation systems, satellite communication, computer networks and innumerable other applications (as shown in Fig.R5-2).

In order to receive radio signals, for instance from AM/FM radio stations, a radio antenna must be used. However, since the antenna will pick up thousands of radio signals at a time, a radio tuner is necessary to tune in to a particular frequency (or frequency range). This is typically done via a resonator (in its simplest form, a circuit with a capacitor and an inductor). The resonator is configured to resonate at a particular frequency (or frequency band), thus amplifying sine waves at that radio frequency, while ignoring other sine waves. Usually, either the inductor or the capacitor of the resonator is adjustable, allowing the user to change the frequency at which it resonates.

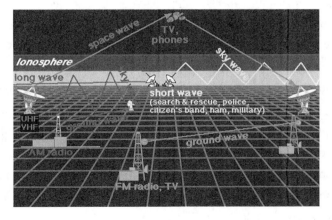

Fig.R5-2 Applications of Radio Waves

6. Power Supply

We use electricity in our homes to perform many different functions. We run electric motors, electric heaters, electric lights and a variety of electronic equipment. Electricity is brought to our homes on electric power lines, and circuits within our homes bring the power to wall receptacles. The electricity that we find at the wall receptacles is a large AC voltage (110 VAC in the U.S. and 220 VAC in P.R.C.) which is suitable for most of our lighting, heating and motor applications, but it is not very suitable for our electronics applications. The electronic circuits in out TVs, VCRs, computers, stereos, etc. all need low level DC voltages. If all we have available in our homes is a large AC voltage, how do we get the low level DC voltage that all of our electronic devices need? The answer is that each of the electronic devices in our homes has a special circuit that converts the AC line voltage to a low level DC voltage. This special circuit is often referred to as a "Power Supply Circuit".

A power supply is a device that produces electricity for use by electronic equipment, or that converts the electricity from the utility mains to a form suitable for use by electronic equipment. Power supplies generally consist either of batteries or of transformer, rectifier, or filter circuits (as shown in Fig.R6-1).

Fig.R6-1 Block Diagrams of Power Supply

A power supply is designed to produce direct current, with alternating-current input. The AC input power is first rectified to provide a unidirectional DC, and then filtered to produce a smooth voltage. Finally, the voltage is regulated to maintain a constant output level despite fluctuations in the power line voltage, current loading or change in temperature.

A direct-current power supply is rated according to its voltage output, its current-delivering capacity, and its ability to maintain a constant voltage under varying load conditions. All of these parameters must be considered when choosing a power supply for use with a given piece of electronic apparatus. The voltage must be correct; the supply must be able to deliver the necessary current; the voltage must remain within a certain range as the load impedance or resistance changes. There is no need, however, to use a supply of much greater precision than required. It would be inefficient, for example, to use a 20-ampere power supply with a circuit that draws only 10 milliamperes.

Power supply voltage can be as small as 1 volt or less, or as large as hundreds of thousands of volts. The current delivering capacity may be just a few milliamperes, or it might be hundreds of amperes. The ability of a power supply to maintain a nearly constant voltage is called the regulation.

Some electronic devices have built-in power supplies. Most radio receivers, tape recorders, hi-fi amplifiers, and other consumer apparatus have built-in power supplies tailored to the requirements of the current. Some equipment, however, requires an external power supply. A wide variety of commercially manufactured power supplies is available for different specialized uses.

Unit II

Electronic Instruments and Products

电子仪器和设备

本单元重点介绍常用的电子仪器和电子设备及其应用,主要包括万用表、数字电压表、示波器、触摸屏、液晶显示器、智能手机、无线电接收机、电视、家庭影院、录像机和数字音响等。通过本单元的学习,学生可以较为系统地学到关于电子技术应用的英语表达,为能在真实电子技术应用专业环境中应用英语进行工作打下基础。

Lesson 7 Multimeter and Its Usage

Multimeters are very useful test instruments. By operating a multi-position switch on the meter they can be quickly and easily set to be a voltmeter, an ammeter or an ohmmeter. They have several settings (called "ranges") for each type of meter and the choice of AC or DC. Some multimeters have additional features such as transistor testing and ranges for measuring capacitance and frequency.

There are two basic types of multimeters, digital and analog. Analog multimeters have a needle and the digital has a LED display.

All digital meters contain a battery to power the display so they use virtually no power from the circuit under test. This means that on their DC voltage ranges they have a very high resistance (usually called input impedance) of 1 MΩ or more, usually 10 MΩ, and they are very unlikely to affect the circuit under test.

Measuring voltage and current with a multimeter

1. Select a range with a maximum greater than you expect the reading to be.
2. Connect the meter, making sure the leads are the correct way round. Digital meters can be

safely connected in reverse, but an analogue meter may be damaged.

3. If the reading goes off the scale: immediately disconnect and select a higher range.

Multimeters are easily damaged by careless use, so please take these precautions.

1. Always disconnect the multimeter before adjusting the range switch.

2. Always check the setting of the range switch before you connect to a circuit.

3. Never leave a multimeter set to a current range (except when actually taking a reading). The greatest risk of damage is on the current ranges because the meter has a low resistance.

Measuring voltage at a point

When testing circuits, you often need to find the voltages at various points, for example the voltage at pin 2 of a 555 timer chip (as shown in Fig.7-1). This can seem confusing—where should you connect the second multimeter lead?

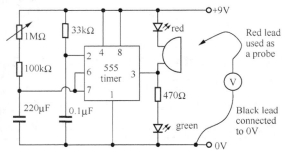

Fig.7-1　Measuring Voltage at a Point

1. Connect the black (negative -) lead to 0 V, normally the negative terminal of the battery or power supply.

2. Connect the red (positive +) lead to the point where you need to measure the voltage.

3. The black lead can be left permanently connected to 0 V while you use the red lead as a probe to measure voltages at various points.

Measuring resistance with a multimeter

Another useful function of the DMM is the ohmmeter. An ohmmeter measures the electrical resistance of a circuit. If you have no resistance in a circuit, the ohmmeter will read 0. If you have an open in a circuit, it will read infinite.

New Words

multimeter ['mʌltimi:tə] n.　　　　　万用表

instrument ['instrəmənt] n.　　　　　工具，仪器，器械

Unit Ⅱ Electronic Instruments and Products 电子仪器和设备

ammeter ['æmi:tə] n.	安培表
ohmmeter ['əummi:tə] n.	欧姆表
needle ['ni:dl] n.	指针
display [di'splei] n.	显示，表现
disconnect [ˌdiskə'nekt] v.	断开
confusing [kən'fju:ziŋ] adj.	使人困惑的，令人费解的
infinite ['infinət] adj.	无限的，无穷的
reverse [ri'və:s] adj.	相反的
pin [pin] n.	大头针，针，拴
precaution [pri'kɔ:ʃn] n.	预防，留心，警戒
risk [risk] n.	风险，危险

Phrases and Expressions

take precautions	采取预防措施
voltage ranges	电压范围
input impedance	输入阻抗

Notes

1. By operating a multi-position switch on the meter they can be quickly and easily set to be a voltmeter, an ammeter or an ohmmeter.

   译文：通过操控万用表上的切换开关，可以快捷而方便地将其设置为电压表、电流表和电阻表。

   by 是介词，因此 by 后面要跟名词、动名词或动词现在分词。

2. …and they are very unlikely to affect the circuit under test.

   译文：它们对被测电路的影响极小。

   be unlikely to：不太可能、不大会。

3. Select a range with a maximum greater than you expect the reading to be.

   译文：它们对被测电路的影响极小。

   reading：读数。

4. When testing circuits, you often need to find the voltages at various points, for example the voltage at pin 2 of a 555 timer chip (as shown in Fig.7-1).

   译文：当我们进行电路测试时，常常需要知道某一点的电压，如图 7-1 中 555 定时器芯片 2 脚的电压。

   在从句和主句有共同的主语时可将从句的主语省略，并将谓语动词改为动词现在分词。从句 When testing circuits 的主语应该是 you，与主句 you often need to find the voltages …中的主语 you 相同，故可省略。

41

Exercises

1. Write T (True) or F (False) beside the following statements about the text.

a. Analog multimeters have a LED display and the digital has a needle.

b. All digital meters contain a electrical motor to power the display so they use virtually no power from the circuit under test.

c. This means that on their DC voltage ranges they have a very high resistance of 10 MΩ or more, usually 100 MΩ, and they are very unlikely to affect the circuit under test.

d. An analogue meter can be safely connected in reverse, but the digital meters may be damaged.

e. Connect the meter, making sure the leads are the correct way round.

f. Always disconnect the multimeter before adjusting the range switch.

2. Fill in the missing words according to the text.

a. By operating a multi-position switch on the meter they can be quickly and easily set to be a _____, an _____ or an _____.

b. They have several settings (called "_____") for each type of meter and the choice of _____ or _____.

c. Some multimeters have additional features such as _____ and ranges for measuring capacitance and _____.

d. There are two basic types of multimeters, _____ and _____.

e. Select a range with a _____ than you expect the reading to be.

f. Multimeters are easily damaged by careless use, so please take the _____.

3. Translate the following paragraph into Chinese.

Measuring resistance with a multimeter

Another useful function of the DMM is the ohmmeter. An ohmmeter measures the electrical resistance of a circuit. If you have no resistance in a circuit, the ohmmeter will read 0. If you have an open in a circuit, it will read infinite.

Chinese Translation of Texts（参考译文）

第 7 课　万用表及其使用

万用表是非常有用的测量仪器，通过操控万用表的切换开关，可以快捷而方便地将其设置为电压表、电流表或电阻表。每种类型的仪表都有几种挡位可以选择（称为量程），还可选择交流挡和直流挡。有些万用表还有一些其他功能，如测量晶体管、测量电容和频率范围等。

万用表有两种类型：数字型和模拟型。模拟万用表用指针显示，数字万用表用液晶显示。

所有的数字仪表都用电池提供显示所需的能量,所以实际上它们并不使用来自被测电路的能量。这就意味着在直流电压测试范围内具有 1 MΩ或更大,通常是 10 MΩ以上的电阻(通常称为输入阻抗),因而它们对被测电路的影响极小。

用万用表测量电压和电流

1. 根据预期的读数范围选择大一级的量程。
2. 连接仪表,确保万用表的表笔以正确的方式形成回路。如果接反了,数字表可能是安全的,模拟表则可能损坏。
3. 如果读数超出刻度范围,应立即断开,并选择更高一级的量程。

万用表使用不当容易损坏,请采取预防措施。

1. 在调节量程前,保持万用表处于断开状态。
2. 在测量前检查量程设置。
3. 绝不把万用表设置在电流挡(除非实际要测电流)。因为万用表的电阻较小,所以在电流挡时最容易损坏。

测量某一点的电压

当我们进行电路测量时,常常需要知道某一点的电压,如图 7-1 中 555 定时器芯片 2 脚的电压。这似乎有点令人迷惑,应该把万用表的表笔连在哪里呢?

1. 将黑表笔接在 0 V,通常是电池或电源的负端。
2. 将红表笔接在需要测量电压的点上。
3. 可以把黑表笔固定在 0 V 位置上,同时用红表笔依次点在各个测量点上测量电压。

用万用表测量电阻

数字万用表的另一个有用的功能是欧姆表。欧姆表是测量电阻的仪表,如果电路没有电阻,则欧姆表读数为 0,如果电路断开,则欧姆表读数为无穷大。

Lesson 8 The Oscilloscope

One of the most important electronic test and measuring instruments is the oscilloscope. It is widely used in design work, trouble-shooting and signal monitoring, manufacturing and production-line testing, and many other applications where the observation of an electrical waveform is desired.

Oscilloscopes range from the basic, or general-purpose types, to the special-purpose types. The general purpose oscilloscope is suitable for any of the usual applications in servicing communications, scientific, or industrial work where the frequency of the signal to be observed does not greatly exceed 500 kHz.

Fig.8-1 is a simple block diagram that shows how an analog oscilloscope displays a measured signal. When you connect an oscilloscope probe to a circuit, the voltage signal travels through the probe to the vertical system of the oscilloscope.

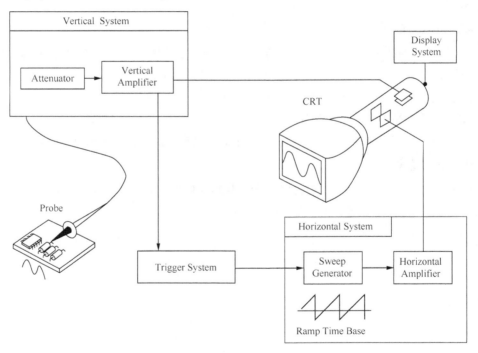

Fig.8-1 Block Diagram of Analog Oscilloscope

The signal travels directly to the vertical deflection plates of the Cathode Ray Tube (CRT). Voltage applied to these deflection plates causes a glowing dot to move. (An electron beam hitting phosphor inside the CRT creates the glowing dot.) A positive voltage causes the dot to move up while a negative voltage causes the dot to move down. Depending on how you set the vertical scale (volts/div control), an attenuator reduces the signal voltage or an amplifier increases the signal voltage.

The signal also travels to the trigger system to start or trigger a "horizontal sweep". Horizontal sweep is a term referring to the action of the horizontal system causing the glowing dot to move across the screen. Triggering the horizontal system causes the horizontal time base to move the glowing dot across the screen from left to right within a specific time interval.

The trigger is necessary to stabilize a repeating signal. It ensures that the sweep begins at the same point of a repeating signal, resulting in a clear picture as shown in Fig.8-2.

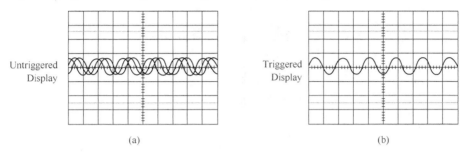

Fig.8-2　Triggering Stabilizes a Repeating waveform

In conclusion, to use an analog oscilloscope, you need to adjust three basic settings to accommodate an incoming signal.

- The attenuation or amplification of the signal. Use the volts/div control to adjust the amplitude of the signal before it is applied to the vertical deflection plates.
- The time base. Use the sec/div control to set the amount of time per division represented horizontally across the screen.
- The triggering of the oscilloscope. Use the trigger level to stabilize a repeating signal.

Also, adjusting the focus and intensity controls enables you to create a sharp, visible display. Fig.8-3 shows an oscilloscope's panel.

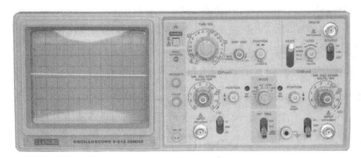

Fig.8-3　Oscilloscope's Panel

New Words

trouble-shoot v.	故障探测
monitor ['mɔnitə] n. & v.	显示器，监视器；监视，监听
attenuator [ə'tenjueitə] n.	衰减器
glowing ['gləuiŋ] adj.	发光的

phosphor ['fɔsfə] n.	荧光粉
trigger ['trigə(r)] n.	触发器
accommodate [ə'kɔmədeit] v.	容纳
incoming ['inkʌmiŋ] adj.	接踵而来的

Phrases and Expressions

production-line	流水线
general-purpose	通用的
special-purpose	专用的
cathode ray tube	阴极射线管
vertical system	垂直系统
vertical deflection plates	垂直偏转板
horizontal system	水平系统
horizontal sweep	水平扫描
ramp time base signal	锯齿波时基信号

Notes

1. It is widely used in design work, trouble-shooting and signal monitoring, manufacturing and production-line testing, and many other applications where the observation of an electrical waveform is desired.

   译文：它被广泛地用于电子设计、故障探测、信号监测、生产制造、流水线检测及许多需要观察电子信号波形的应用场合。

   where 此处不是指地点，而是指"领域"或"应用场合"。

2. Oscilloscopes range from the basic, or general-purpose types, to the special-purpose type.

   译文：示波器的类型包括通用示波器和专用示波器。

   range from: 从……到……变动或从……到……范围。

   例如，The children's ages range from 8 to 15.

   译文：这些孩子们的年龄在 8 岁到 15 岁之间。

3. The general purpose oscilloscope is suitable for any of the usual applications in servicing communications, scientific, or industrial work where the frequency of the signal to be observed does not greatly exceed 500 kHz.

   译文：通用示波器适用于通信、科学研究、工业生产等各个领域，用来观察频率不超过 500 kHz 的电子信号。

   in: 在……方面或在……领域。

Unit Ⅱ Electronic Instruments and Products 电子仪器和设备

4. Horizontal sweep is a term referring to the action of the horizontal system causing the glowing dot to move across the screen.

译文：水平扫描是指水平系统使光点沿水平方向扫过屏幕。

term: 术语，而 referring 是动词现在分词，作后置定语。

5. The trigger is necessary to stabilize a repeating signal. It ensures that the sweep begins at the same point of a repeating signal, resulting in a clear picture as shown in Fig. 8-2.

译文：稳定重复显示的波形需要触发。它保证重复显示波形时，从同一点开始扫描，使波形如图 8-2 所示清晰地显示。

resulting in…：是动词现在分词作伴随状语。

resulting in…：导致，表结果。

例如：Sometimes I pay, sometimes he pays—it seems to average out (result in a fair balance) in the end.

译文：有时我付钱，有时他付钱——到头来似乎两相抵销。

Exercises

1. Write T (True) or F (False) beside the following statements about the text.

a. The general purpose oscilloscope is suitable for special-purpose applications.

b. The general purpose oscilloscope is used in servicing communications, scientific, or industrial work where the frequency of the signal to be observed does not greatly exceed 1000 kHz.

c. The signal travels directly to the vertical deflection plates of the Cathode Ray Tube (CRT).

d. Voltage applied to these deflection plates causes a glowing dot to less.

e. A positive voltage causes the dot to move down while a negative voltage causes the dot to move up.

2. Fill in the missing words according to the text.

a. One of the most important _____ and _____ is the oscilloscope.

b. It is widely used in design work, _____ and _____, manufacturing and production-line testing, and many other applications where the observation of an electrical _____ is desired.

c. Oscilloscopes range from the _____, to the special-purpose types.

d. When you connect an oscilloscope probe to a circuit, the _____ travels through the probe to the vertical system of the _____.

e. An electron beam hitting phosphor inside the _____ creates the glowing dot.

f. The signal also travels to the trigger system to start or trigger a "_____".

3. Translate the following paragraphs into Chinese.

In conclusion, to use an analog oscilloscope, you need to adjust three basic settings to accommodate an incoming signal:

- The attenuation or amplification of the signal. Use the volts/div control to adjust the amplitude

47

of the signal before it is applied to the vertical deflection plates.
- The time base. Use the sec/div control to set the amount of time per division represented horizontally across the screen.
- The triggering of the oscilloscope. Use the trigger level to stabilize a repeating signal.

Also, adjusting the focus and intensity controls enables you to create a sharp, visible display.

Chinese Translation of Texts（参考译文）

第 8 课　示波器

示波器是最重要的电子测量仪器之一。它被广泛地用于电子设计、故障探测、信号监测、生产制造、流水线检测及许多需要观察电子信号波形的应用场合。

示波器的类型包括通用示波器和专用示波器。通用示波器适用于通信、科学研究、工业生产等各个领域，用来观察频率不超过 500 kHz 的电子信号。

图 8-1 是一个简单的方框图，表示了模拟示波器是如何显示被测信号的。当你把示波器探头连上电路时，电压信号就通过探头进入示波器的垂直系统。

信号直接进入阴极射线管（CRT）的垂直偏转板。加在这个偏转板上的电压使光点运动。（电子束撞击 CRT 内的荧光屏产生光点。）正电压使光点向上运动，而负电压使光点向下运动。衰减器减小显示的信号电压或放大器增大显示的信号电压，取决于你如何调节垂直偏转旋钮（volts/div 控制旋钮）。

信号还进入触发系统开始或触发"水平扫描"。水平扫描是指水平系统使光点沿水平方向扫过屏幕。对水平系统的触发使水平时基电路在专门的时间间隔内让光点从左至右扫过屏幕。

稳定重复显示的波形需要触发系统。它保证重复显示波形时，从同一点开始扫描，使波形如图 8-2 所示清晰地显示。

总之，使用模拟示波器时，为了完全显示连续出现的信号波形，需要调节 3 个基本按键。
- 衰减或增大信号：使用"volts/div"控制键，以便在信号还未加到垂直偏转板之前调节信号的振幅。
- 时基：使用"sec/div"控制键，设置屏幕水平方向每格所代表的时间长短。
- 示波器的触发：使用"触发电平"控制键稳定重复显示的信号。

还可以调节聚焦和亮度控制旋钮，以保证显示的波形尖锐和可见。图 8-3 所示为示波器的面板图。

Unit II Electronic Instruments and Products 电子仪器和设备

Lesson 9 Virtual Instruments

Virtual instruments are computer programs that interact with real world objects by means of sensors and implement functions of real or imaginary instruments. The sensor is usually a simple hardware that acquires data from the object, transforms it into electric signals and transmits it to the computer for further processing (as shown in Fig.9-1). Simple virtual measuring instruments just acquire and analyse data, but more complex virtual instruments communicate with object in both directions.

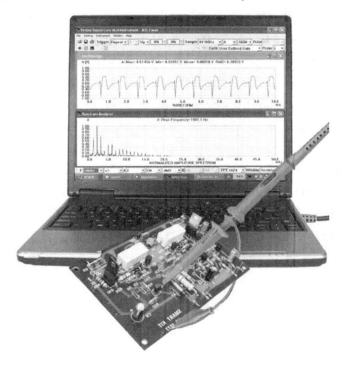

Fig.9-1 Virtual Instrument

Real world signals are of analogue nature, while a computer is a digital instrument; therefore the computer needs also interpreters—analogue-to-digital and digital-to-analogue converters for communication with the object. ADC and DAC boards that implement this function in inexpensive systems are usually placed inside the computer. Compact external ADC/DAC converters with USB interface are also becoming popular.

LabVIEW is a graphical programming language from National Instruments. LabVIEW programs are called virtual instruments often abbreviated to VIs. Each virtual instrument has two components, a block diagram and a front panel. Controls and indicators on the front panel allow an operator to input data into or extract data from an already running virtual instrument.

In terms of performance, LabVIEW includes an actual compiler that produces native code for the CPU platform, so the graphical code is compiled, rather than interpreted.

One of the main benefits of LabVIEW is that people with little or no previous programming experience are able to access hardware input/output more rapidly and through a hardware abstraction system. This abstraction allows isolation between hardware implementation and software solution. A technique which without National Instruments software driver interface would be extremely time consuming.

Another virtual instrumentation component is modular I/O, designed to be rapidly combined in any order or quantity to ensure that virtual instrumentation can both monitor and control any development aspect. Using well-designed software drivers for modular I/O, engineers and scientists can quickly access functions during concurrent operation.

The third virtual instrumentation element—using commercial platforms, often enhanced with accurate synchronization—ensures that virtual instrumentation takes advantage of the very latest computer capabilities and data transfer technologies. This element delivers virtual instrumentation on a long-term technology base that scales with the high investments made in processors, buses, and more.

Virtual instrumentation systems frequently use Ethernet for remote test system control, distributed I/O, and enterprise data sharing. Ethernet provides a low-cost, moderate-throughput method for exchanging data and control commands over distances. However, due to its packet-based architecture, Ethernet is not deterministic and has relatively high latency. For some applications, such as instrumentation systems, the lack of determinism and high latency make Ethernet a poor choice for integrating adjacent I/O modules. These situations are better served with a dedicated bus such as PXI, VXI, or GPIB（as shown in Fig.9-2）.

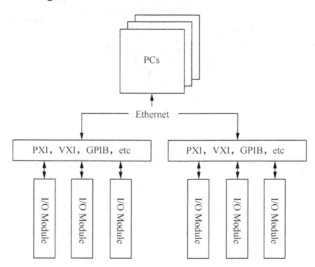

Fig.9-2 Virtual instrumentation systems are served with a dedicated bus

New Words

virtual ['və:tʃuəl] adj.　　　　　　　　虚拟的

hardware ['hɑ:dweə] n.　　　　　　　　计算机硬件

transform [træns'fɔ:m] v.	改变，改观，变换
transmit [trænz'mit] v.	传输，传送，发送
interpreter [in'tə:pritə] n.	解释者，翻译器
converter [kən'və:tə] n.	变换器，换流器，变压器，变频器
implementation [ˌimplimen'teiʃn] n.	成就，贯彻，安装启用
modular ['mɔdjələ] adj.	模块化的
monitor ['mɔnitə] n. & vt.	显示屏，屏幕，显示器，监测仪；监视
synchronization [ˌsiŋkrənai'zeiʃən] n.	同一时刻，同步
bus [bʌs] n.	总线
Ethernet ['i:θənet] n.	以太网
latency ['leitənsi] n.	潜伏，潜在因素
integrate ['intigreit] v.	使一体化，使集成
instrumentation [ˌinstrəmen'teiʃən] n.	仪器

Phrases and Expressions

ADC（Analogue-to-Digital Converter）	模数转换器
DAC（Digital-to-Analogue Converter）	数模转换器
USB interface	USB 接口
CPU（Central Processing Unit）	中央处理单元
scale with	与……成比例

Notes

1. Virtual instruments are computer programs that interact with real world objects by means of sensors and implement functions of real or imaginary instruments.

   译文：虚拟仪器是通过传感器与真实世界的物体进行交互并完成真实或虚拟仪器功能的计算机程序。

   real world objects: 真实世界的物体，现实世界中的物体。

2. Real world signals are of analogue nature, while a computer is a digital instrument.

   译文：现实世界的信号是模拟信号，而计算机是数字仪器。

   real world signals: 现实世界的信号。

3. therefore the computer needs also interpreters—analogue-to-digital and digital-to-analogue converters for communication with the object.

   译文：因此计算机与对象进行通信时需要进行转换——模数转换和数模转换，与被测对象进行通信。

   analogue-to-digital: 模拟-数字，从模拟到数字。

   digital-to-analogue converter: 数字-模拟转换器。

4. LabVIEW is a graphical programming language from National Instruments. LabVIEW programs are called virtual instruments often abbreviated to VIs.

译文：LabVIEW 是 NI 公司的图形化编程语言。通常把 LabVIEW 程序称为虚拟仪器，缩写为 VIs。

LabVIEW：美国国家仪器公司（National Instruments，NI）1986 年推出的图形化系统设计软件，现今，数以百万的工程师和科学家可以使用 LabVIEW 来构建他们的测试、测量与控制系统。

VIs：virtual instrument，虚拟仪器技术是利用高性能的模块化硬件，结合高效灵活的软件来完成各种测试、测量和自动化的应用。

5. Another virtual instrumentation component is modular I/O.

译文：另一个虚拟仪器组件是模块化的 I/O。

I/O：input/output 的缩写，即输入/输出端口。

6. These situations are better served with a dedicated bus such as PXI, VXI, or GPIB.

译文：这种情况下最好选择专用总线，如 PXI、VXI 或 GPIB。

PXI：PCI Extensions for Instrumentation，面向仪器系统的 PCI 扩展，是一种坚固的基于 PC 的测量和自动化平台。

VXI：Cisco Virtualization Experience Infrastructure，虚拟化体验基础架构。

GPIB：General Purpose Interface Bus，通用接口总线。

Exercises

1. Write T (True) or F (False) beside the following statements about the text.

a. The sensor is usually a simple software that acquires data from the object, transforms it into electric signals and transmits it to the computer for further processing.

b. Real world signals are of digital nature, while a computer is a analogue instrument.

c. ADC and DAC boards that implement this function in expensive systems are usually placed inside the computer.

d. LabVIEW is a graphical programming language from National Instruments.

e. Controls and indicators on the front panel allow an operator to input data into or extract data from an already running virtual instrument.

2. Fill in the missing words according to the text.

a. Virtual instruments are computer programs that interact with real world objects by means of sensors and implement functions of _____.

b. LabVIEW programs are called _____ often abbreviated to VIs.

c. This abstraction allows isolation between hardware implementation and _____.

d. Ethernet provides _____, _____ method for exchanging data and control commands over distances.

e. This element delivers _____ on a long-term technology base that scales with the high investments made in ＿＿＿＿＿＿, ＿＿＿＿＿＿, and more.

3. Translate the following sentences into Chinese.

a. Simple virtual measuring instruments just acquire and analyse data, but more complex virtual instruments communicate with objects in both directions.

b. In terms of performance, LabVIEW includes an actual compiler that produces native code for the CPU platform, so the graphical code is compiled, rather than interpreted.

Chinese Translation of Texts（参考译文）

第9课　虚　拟　仪　器

　　虚拟仪器是通过传感器与真实世界的物体进行交互并完成真实或虚拟仪器功能的计算机程序。传感器通常是一个简单的硬件，它从实物获取数据，将其转换成电子信号并传输到计算机作进一步处理（见图9-1）。简单的虚拟测量仪器只能获取和分析数据，但更复杂的虚拟仪器能与物体进行双向通信。

　　现实世界的信号是模拟信号，而计算机是数字仪器；因此计算机与对象进行通信时需要进行转换——模数和数模转换。在廉价的系统中，实现这一功能的模-数转换和数-模转换板通常内置于计算机中。紧凑的外置带USB接口的模数/数模转换器也越来越普及。

　　LabVIEW是NI公司的图形化编程语言。通常把LabVIEW程序称为虚拟仪器，缩写为VIs。每个虚拟仪器有两个组成部分：框图和前面板。前面板上的控制和指令允许操作员从一个已经在运行的虚拟仪器中输入数据或提取数据。

　　在性能方面，LabVIEW包含一个为CPU平台生成本地代码的实际的编译器，它编译图形代码，而不仅仅是转换。

　　LabVIEW的主要优点之一是，有很少的或者根本没有编程经验的人都能够很迅速地访问硬件的输入/输出，以及通过硬件抽象系统。这个硬件抽象层允许硬件实现和软件解决方案之间的隔离。没有NI软件驱动程序接口的技术将是非常费时的。

　　另一个虚拟仪器组件是模块化的I/O，它被设计用来在任何命令或数量上迅速兼容，以确保虚拟仪器可以同时监视和控制各个方面的发展。使用为模块化的I/O精心设计的软件驱动程序，工程师和科学家可以在并行操作期间实现快速访问功能。

　　虚拟仪器的第三个要素——使用商业平台，这通常使得精确度同步得到提高——确保该虚拟仪器采用了最新的计算机功能和数据传输技术。这个元素保证了虚拟仪器在高投资规模的处理器、总线等技术上的技术水平。

　　虚拟仪器系统在进行远程测试系统控制、分布式I/O和企业数据共享时，经常利用以太网。以太网为数据交换和远距离指令控制提供了一个低成本和中等吞吐量的方法。然而，由于其基于分组的架构，以太网具有不确定性，并有较高的延迟。对于某些应用，如仪表系统，由于以太网缺乏确定性和高延迟，使其在集成相邻的I/O模块时不是一种好的选择。这种情况下最好选择专用总线，如PXI、VXI或GPIB（见图9-2）。

Lesson 10 Portable Media Player

A Portable Media Player (PMP) is a consumer electronics device that is capable of storing and playing digital media such as audio, images, video, documents, etc. The data is typically stored on a hard drive, microdrive, or flash memory (as shown in Fig.10-1).

Fig.10-1 Portable Media Players

Digital audio players are generally categorized by storage media.

Flash-based players: These are non-mechanical solid state devices that hold digital audio files on internal flash memory. Because they are solid state and do not have moving parts, they require less battery power, and may be more resilient to hazards such as dropping or fragmentation than hard disk-based players.

Hard drive-based players or digital jukeboxes: Devices that read digital audio files from a hard disk drive (HDD). At typical encoding rates, this means that tens of thousands of songs can be stored on one player. The disadvantages with these units are that a hard drive consumes more power, is larger and heavier and is inherently more fragile than solid-state storage.

MP3 CD/DVD players: Portable CD players that can decode and play MP3 audio files stored on CDs. Such players are typically much less expensive than either the hard drive or flash-based players. The blank CD-R media they use is very inexpensive, typically costing less than US$0.15 per disc. These devices have the feature of being able to play standard "Red book" CD-DA audio CDs. A disadvantage is that due to the low rotational disk speed of these devices, they are even more susceptible to skipping or other misreads of the file if they are subjected to uneven acceleration (shaking) during playback. The mechanics of the player itself "however" can be quite sturdy, and are generally not as prone to permanent damage due to being dropped as hard drive-based players. Since a CD can typically hold only around 700 megabytes of data, a large library will require multiple disks to contain. However, some higher-end units are also capable of reading and playing back files stored on larger capacity DVD; some also have the ability to play back and display video content, such as movies. An additional consideration can be the relatively large width of these devices, since they have to be able to fit a CD.

Networked audio players: Players that connect via (Wi-Fi) network to receive and play audio. These types of units typically do not have any local storage of their own and must rely on a server, typically a personal computer also on the same network, to provide the audio files for playback.

USB host/memory card audio players: Players that rely on USB flash drives or other memory cards to read data.

PMPs are capable of playing digital audio, images, and video. Usually, a color Liquid Crystal Display (LCD) or Organic Light-Emitting Diode (OLED) screen is used as a display. Various players include the ability to record video, usually with the aid of optional accessories or cables, and audio, with a built-in microphone or from a line out cable or FM tuner. Some players include readers for memory cards.

New Words

store [stɔ:] v.	存储
audio ['ɔ:diəu] n.	音频
categorize ['kætəgəraiz] v.	分类
storage ['stɔ:ridʒ] n.	存储，仓库
solid ['sɔlid] adj.	固体的
skip [skip] v.	跳读
resilient [ri'ziliənt] adj.	有弹性的
hazard ['hæzəd] n.	危险
fragmentation [frægmen'teiʃn] n.	碎裂
jukebox ['dʒu:kbɔks] n.	自动点唱机
capacity [kə'pæsəti] n.	容量
encode [in'kəud] v.	编码
inherently [in'hiərəntli] adv.	内在地，固有地
fragile ['frædʒail] adj.	脆弱的，易碎的
decode [di:'kəud] v.	解码
rotational [rəu'teiʃənl] adj.	转动的
susceptible [sə'septəbl] adj.	易受……影响的
sturdy ['stə:di] adj.	结实的，坚固的
megabyte ['megəbait] n.	兆字节
server ['sə:və] n.	服务器
accessory [ək'sesəri] n.	配件，附件
cable ['keibl] n.	电缆

Phrases and Expressions

PMP (Portable Media Player)　　便携式媒体播放器
HDD (Hard Disk Drive)　　硬盘驱动器
CD-DA (Compact Disc Digital Audio)　　数字音频光碟
LCD (Liquid Crystal Display)　　液晶显示器
OLED (Organic Light-emitting Diode)　　有机发光二极管

Notes

1. Because they are solid state and do not have moving parts, they require less battery power, are less likely to skip during playback, and may be more resilient to hazards such as dropping or fragmentation than hard disk-based players.

   译文：因为它们是静态存储装置，并没有移动的部件，因而需要的电量更少，在播放过程中很少会出现跳读现象，而且与硬盘存储播放器相比，更能经受住诸如掉落或碎裂等此类情况所造成的损害。

2. …they are even more susceptible to skipping or other misreads of the file if they are subjected to uneven acceleration (shaking) during playback.

   译文：如果它们在播放过程中遇到不规则的加速度（震动），就更容易出现跳读或误读文件的情况。

   be subjected to: 遭受，经受，如 I'd rather not live in an area that is subject to flooding.

3. The mechanics of the player itself however can be quite sturdy, and are generally not as prone to permanent damage due to being dropped as hard drive-based players.

   译文：这种播放器比硬盘型播放器坚固，一般不容易因摔到地上而出现永久性损坏。

   as … as：指"两个人或物在某方面一样"，既可以用在肯定句中，也可以用在否定句中，而 so … as …只能用在否定句里。

   prone to: 很可能，如 People are more prone to make mistakes when they are tired.

Exercises

1. Write T (True) or F (False) beside the following statements about the text.

a. A PMP is a device that is only capable of playing digital audio and video.

b. Flash-based players are consumer electronics which are solid state and have moving parts.

c. Compared to a hard drive, solid-state storage is smaller and lighter.

d. A CD can hold as much data as you want.

e. CD players are likely to skip or misread of the file during playback if mishandled.

f. Networked audio players usually have to depend on a server as they do not have any local storage of their own.

g. With the help of some accessories or cable, a PMP can record video as well as audio.

2. Fill in the missing words according to the text.

a. Digital audio players are generally categorized by such _____ media as _____, _____, _____, _____, _____.

b. At typical _____ rates, this means that tens of thousands of songs can be _____ on one player.

c. A hard drive which _____ more power is larger and heavier, and is inherently more than solid-state storage.

d. Networked audio players are the players that connect _____ to receive and play audio.

3. Translate the following paragraph into Chinese.

Hard drive-based players or digital jukeboxes: Devices that read digital audio files from a Hard Disk Drive (HDD). These players have higher capacities as of 2010 ranging up to 500 GB. At typical encoding rates, this means that tens of thousands of songs can be stored on one player. The disadvantages with these units are that a hard drive consumes more power, is larger and heavier and is inherently more fragile than solid-state storage.

Chinese Translation of Texts（参考译文）

第 10 课　便携式媒体播放器

便携式媒体播放器（PMP）是一个消费性电子装置，能够存储和播放数字媒体，如音频、图像、视频和文件等。数据通常存储在硬盘驱动器、微驱动器或闪存中。

数字音频播放器通常按照存储介质来分类。

基于 Flash 的播放器：数字音频文件储存在内部闪存中的非机械固态设备。因为它们是静态存储装置，并没有移动的部件，因而需要的电量更少，在播放过程中很少会出现跳读现象，而且与硬盘存储播放器相比，更能经受住诸如掉落或碎裂等此类情况所造成的损害。

基于硬盘的播放器或数字点唱机：这种播放器从硬盘驱动器（HDD）读取数字音频文件。按照典型的编码率，这意味着上万首歌曲可以存储在一个播放器里。其缺点是硬盘驱动器消耗更多的电力，播放器较大和较重，比固态存储更容易损坏。

MP3 CD/DVD 播放机：便携式 CD 播放机可以解码和播放存储在光盘上的 MP3 音频文件。这样的播放器通常比硬盘或闪存播放器便宜得多。它们使用的空白的 CD-R 非常便宜，一般每张光盘花费不到 0.15 美元。这些器件具有能够播放标准的"红皮书"CD-DA 音频 CD 的功能。其缺点是，由于这些设备的旋转磁盘速度很低，如果它们在播放过程中遇到不规则的加速度（震动），就更容易出现跳读或误读文件的情况。这种播放器比硬盘型播放器坚固，一般不容易因摔到地上而出现永久性损坏。由于 CD 通常只能容纳大约 700 兆字节的数据，因此一个大型文库往往需要多个磁盘来容纳。然而，一些较高端的播放器也能读取和播放那些存储在容量更大的 DVD 中的文件；有的播放器还能播放和显示视频内容，如电影。我们还需要了解的是，这些设备具有较大的宽度，因为它们要容纳整个 CD 光盘。

　　网络音频播放器：播放器通过联网（Wi-Fi）接收和播放音频。这种类型的播放器一般没有自己的存储器，需要依靠一台服务器（通常为一台连接在同网络上的私人计算机）来提供音频文件进行播放。

　　USB 主机/存储卡音频播放器：播放器依靠 USB 闪存驱动器或其他存储卡来读取数据。

　　PMP 能够播放数字音频、图像和视频。一般使用彩色液晶显示屏（LCD）或有机发光二极管（OLED）屏幕进行显示。不同类型的播放器包括的功能为，录制视频，通常借助可选附件或电缆；录制音频，借助一个内置麦克风或线路输出电缆或 FM 调谐器。一些播放器具有存储卡读取器。

Unit II Electronic Instruments and Products 电子仪器和设备

Lesson 11 LCD (Liquid Crystal Display)

LCDs are used in a wide range of applications, including computer monitors, television, instrument panels, aircraft cockpit displays, signage, etc. They are common in consumer devices such as video players, gaming devices, clocks, watches, calculators, and telephones. LCDs have displaced Cathode Ray Tube (CRT) displays in most applications. They are usually more compact, lightweight, portable, less expensive, more reliable, and easier on the eyes. They are available in a wider range of screen sizes than CRT and plasma displays.

A liquid crystal display (as shown in Fig.11-1) is a thin, flat electronic visual display that uses the light modulating properties of Liquid Crystals (LCs). LCs do not emit light directly.

One feature of liquid crystals is that they're affected by electric current. A particular sort of nematic liquid crystal, called Twisted Nematics (TN), is naturally twisted. Applying an electric current to these liquid crystals will untwist them to varying degrees, depending on the current's voltage.

Fig.11-1 LCD (Liquid Crystal Display)

There's far more to building an LCD than simply creating a sheet of liquid crystals. The combination of four facts makes LCD possible (as Fig.11-2).

- Light can be polarized.
- Liquid crystals can transmit and change polarized light.
- The structure of liquid crystals can be changed by electric current.
- There are transparent substances that can conduct electricity.

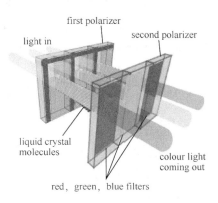

Fig.11-2 How does LCD work

An LCD is a device that uses these four facts in a surprising way!

But just what are these things called liquid crystals? The name "liquid crystal" sounds like a

59

contradiction. We think of a crystal as a solid material like quartz, usually as hard as rock, and a liquid is obviously different. How could any material combine the two?

We learned in school that there are three common states of matter: solid, liquid or gaseous. Solids act the way they do because their molecules always maintain their orientation and stay in the same position with respect to one another. The molecules in liquids are just the opposite: they can change their orientation and move anywhere in the liquid. But there are some substances that can exist in an odd state that is sort of like a liquid and sort of like a solid. When they are in this state, their molecules tend to maintain their orientation, like the molecules in a solid, but also move around to different positions, like the molecules in a liquid. This means that liquid crystals are neither a solid nor a liquid. That's how they ended up with their seemingly contradictory name.

So, do liquid crystals act like solids or liquids or something else? It turns out that liquid crystals are closer to a liquid state than a solid. It takes a fair amount of heat to change a suitable substance from a solid into a liquid crystal, and it only takes a little more heat to turn that same liquid crystal into a real liquid. This explains why liquid crystals are very sensitive to temperature, and why they are used to make thermometers and mood rings. It also explains why a laptop computer display may act funny in cold weather or during a hot day at the beach.

There are two main types of LCDs used in computers: passive matrix and active matrix.

Most LCD displays use active matrix technology. A Thin Film Transistor (TFT) arranges tiny transistors and capacitors in a matrix on the glass of the display. To address a particular pixel, the proper row is switched on, and then a charge is sent down the correct column. Since all of the other rows that the column intersects are turned off, only the capacitor at the designated pixel receives a charge. The capacitor is able to hold the charge until the next refresh cycle.

The other type of LCD technology is passive matrix. This type of LCD display uses a grid of conductive metal to charge each pixel. Although they are less expensive to produce, passive matrix monitors are rarely used today due to the technology's slow response time and imprecise voltage control compared to active matrix technology.

New Words

cockpit ['kɔkpit] n. （飞机驾驶员的）驾驶舱，座舱
signage ['sainidʒ] n. 引导标示
plasma ['plæzmə] n. 等离子体
laptop ['læptɔp] n. 笔记本电脑
nematic [ni'mætik] adj. （液晶）【晶体】向列的
polarize ['pəuləraiz] vt. （使）偏振，（使）极化
transparent [træns'pærənt] adj. 透明的，显然的

contradiction [ˌkɔntrə'dikʃn] n.	矛盾，否认，反驳
quartz [kwɔ:ts] n.	石英
gaseous ['gæsiəs] adj.	气态的，气体的
molecule ['mɔlikju:l] n.	分子，微小颗粒，微粒
orientation [ˌɔ:riən'teiʃn] n.	方向，定向
thermometer [θə'mɔmitə] n.	温度计，体温计
matrix ['meitriks] n.	【数】矩阵，模型
transistor [træn'zistə] n.	晶体管
capacitor [kə'pæsitə] n.	电容器
pixel ['piks(ə)l] n.	（显示器或电视机图像的）像素（等于 picture element）
intersect [ˌintə'sekt] vt. & vi.	横断，横切，贯穿；相交，交叉
designate ['dezigneit] vt.	指定，指派
grid [grid] n.	格子，网格

Phrases and Expressions

LCD (Liquid Crystal Display)	液晶显示器
CRT (Cathode Ray Tube)	阴极射线管
TN (Twisted Nematics)	扭曲向列型
TFT (Thin Film Transistor)	薄膜晶体管
refresh cycle	恢复周期，【计】刷新周期，更新周期

Notes

1. Applying an electric current to these liquid crystals will untwist them to varying degrees, depending on the current's voltage.

   译文：在这种液晶中有电流流过时，液晶会根据所加的电压改变其扭曲程度。

   句中 Applying an electric current to these liquid crystals 是动名词短语作主语。英语中，在动词的基础上加-ing，可使该动词或动词短语有名词的各种特征，可作名词灵活使用。如 Climbing mountains is really fun.

2. There's far more to building an LCD than simply creating a sheet of liquid crystals.

   译文：并不是简单地铺一层液晶就可以做成LCD。

   far: 副词，非常，例如，(1) There are far more opportunities for young people than there used to be. (2) He always gives us far too much homework.

3. Solids act the way they do because their molecules always maintain their orientation and stay in the same position with respect to one another.

译文：固体之所以表现为这样是因为，它们的分子总是保持自己的方向，并保持相对位置不变。

with respect to：关于，相对于。

4. That's how they ended up with their seemingly contradictory name.

译文：这就是最终它们有一个看似矛盾的名字的原因。

end up (with)：以……告终，例如，(1) If you go on like this you will end up with nothing. (2) I end up doing all the work myself. (3) If you go on like this, you will end up in prison.

5. It takes a fair amount of heat to change a suitable substance from a solid into a liquid crystal, and it only takes a little more heat to turn that same liquid crystal into a real liquid.

译文：将一个合适的物质从固体转换为液晶需要相当多的热量，而将同样的液晶变成真正的液体只需要稍微多一点的热量。

(1) fair：（数量，尺寸，程度，距离等）相当大的，例如，My birthday is still fair way off.

(2) it takes … to：需要（某种品质或事物）来做某事，例如，It takes courage to say what you think.

Exercises

1. Write T (True) or F (False) beside the following statements about the text.

a. LCDs are widely used in common consumer devices.

b. Liquid crystals can emit light directly in a LCD.

c. Electric current can change the structure of liquid crystals.

d. Molecules always maintain their orientation in liquids.

e. Molecules in crystal liquids can not only maintain their orientation but move around to different positions.

f. As liquid crystals are very sensitive to temperature, they can be used to make thermometers.

g. Because of the slow response time and imprecise voltage control, passive matrix monitors are rarely used today.

2. Fill in the missing words according to the text.

a. A liquid crystal display is a thin, flat _____ that uses _____ of liquid crystals.

b. Liquid crystals can _____ and _____ polarized light.

c. There are three common states of matter: _____.

d. The molecules in liquids can change their _____ and move anywhere.

e. _____ and _____ are two main types of LCDs used in computers.

3. Translate the following paragraph into Chinese.

Solids act the way they do because their molecules always maintain their orientation and stay in the same position with respect to one another. The molecules in liquids are just the opposite: they can change their orientation and move anywhere in the liquid. But there are some substances that can exist in an odd state that is sort of like a liquid and sort of like a solid. When they are in this state, their molecules tend to maintain their orientation, like the molecules in a solid, but also move around to different positions, like the molecules in a liquid. This means that liquid crystals are neither a solid nor a liquid.

Chinese Translation of Texts（参考译文）

第 11 课 液晶显示器

液晶显示器有广泛的应用场景，包括计算机显示器、电视、仪表面板、飞机座舱显示器和招牌等。它们是常见的消费设备，如视频播放器、游戏设备、钟、手表、计算器和电话。液晶显示器在大多数应用中已经取代了阴极射线管（CRT）显示器。它们通常更紧凑，重量轻，便于携带，更便宜，更可靠，对眼睛来说更舒服。比起 CRT 和等离子显示器，它们可以适应更广泛的屏幕尺寸。

液晶显示器（LCD）是一种平面薄型视觉显示器，采用液晶的光调制特性。液晶不直接发光。液晶的一个特点是液晶容易受电流影响。一种被称为扭曲向列的特殊向列液晶，其本身是扭曲的，在这种液晶中有电流流过时，液晶会根据所加的电压改变其扭曲程度。

并不是简单地铺一层液晶就可以做成 LCD，做成一个 LCD 要结合以下 4 点。

- 偏振光的产生。
- 液晶可以传输和改变偏振光。
- 液晶的结构可以用电流来改变。
- 有可以导电的透明物质。

LCD 就是奇妙地结合了这 4 种技术的产品。

但这些被称为液晶的东西是什么？液晶这个名字听起来好像很矛盾。我们认为，水晶是像石英一样的固体物质，通常硬如石头，而液体显然是不同的。如何合并两种材料呢？

我们在学校里学过，物质有 3 种常见的状态：固体、液体或气体。固体之所以表现为这样是因为，它们的分子总是保持自己的方向，并保持相对位置不变。液体中的分子则正好相反：在液体里它们可以改变方向并向任何地方移动。但也有一些物质，它们可以以一个奇怪的状态存在，这种状态有点儿像液体，又有点像固体。当它们在这种状态下时，它们的分子就倾向于保持自己的方向，就像一个固体的分子，但也能到处移动到不同的位置，就像液体中的分子。这意味着液晶既不是固体也不是液体。这就是最终它们有一个看似矛盾的名字的原因。

那么，液晶表现得更像固体还是液体，抑或是其他什么东西？原来，比起固体来液晶更接近于液体。将一个合适的物质从固体转换为液晶需要相当多的热量，而将同样的液晶变成真正的液体只需要稍微多一点的热量。这就解释了为什么液晶对温度非常敏感，它们为什么可以用来制作温度计和情绪戒指。这也解释了为什么一个笔记本电脑显示器会在寒冷的天气里或在热天的海滩上表现怪异。

计算机所用的 LCD 主要有两种类型：无源矩阵和有源矩阵。

大部分 LCD 显示器用有源矩阵技术，在显示器的玻璃上铺了一层以矩阵形式排列的小晶体管和电容构成的晶体管薄膜（TFT），为定位某个像素，相应行的开关会合上，然后把电荷送到相应的列上。该列上其他点的行开关都断开了，只有指定像素点的电容可以接收电荷。电容可以保持这个电荷直到下一次刷新。

另一种 LCD 技术是无源矩阵技术，这种 LCD 显示器用一个导电金属栅给各像素点充电，虽然制造成本比较低，但现在无源矩阵很少用，因为与有源矩阵技术相比这种技术的时间响应慢，且电压控制不够精确。

Unit Ⅱ Electronic Instruments and Products 电子仪器和设备

Lesson 12   Smartphone

You probably hear the term "Smartphone" tossed around a lot. But if you've ever wondered exactly what a Smartphone is, well, you're not alone. How is a Smartphone different than a cell phone, and what makes it so smart?

In a nutshell, a Smartphone is a device that lets you make telephone calls, but also adds in features that, in the past, you would have found only on a personal digital assistant or a computer, such as the ability to send and receive E-mail and edit Office documents.

However, to really understand what a Smartphone is (and is not), we should start with a history lesson. In the beginning, there were cell phones and personal digital assistants (or PDAs). Cell phones were used for making calls—and not much else—while PDAs, like the Palm Pilot, were used as personal, portable organizers. A PDA could store your contact info and a to-do list, and could SYNC with your computer.

PDAs gained wireless connectivity and were able to send and receive E-mail. Cell phones, meanwhile, gained messaging capabilities, too. PDAs then added cellular phone features, while cell phones added more PDA-like (and even computer-like) features. The result was the Smartphone.

Unlike many traditional cell phones, Smartphones allow individual users to install, configure and run applications of their choosing (as shown in Fig.12-1). A Smartphone offers the ability to conform the device to your particular way of doing things. Most standard cell-phone software offers only limited choices for re-configuration, forcing you to adapt to the way it's set up. On a standard phone, whether or not you like the built-in calendar application, you are stuck with it except for a few minor tweaks. If that phone were a Smartphone, you could install any compatible calendar application you like. For example, a Smartphone will enable you to do more.

Fig.12-1   Smartphone

It may allow you to create and edit Microsoft documents or at least view the files. It may allow you to download apps as well.

Smartphones are run by using Operating Systems just like a computer does, some of the more popular operating systems or OS are the Android, iOS, Windows, Symbian and Blackberry OS.

The Android OS was released in 2008 and it is considered an open source platform that was supported by Google. The features it included were Maps, Calendar, Gmail and a fully functioning HTML web browser. One of the extremely popular Smartphone is the iPhone from Apple. It was first introduced to the world in 2007 and it priced at a high $ 499. The iPhone was the first phone to have a large touch screen that was designed for direct finger input.

Since cell phones and PDAs are the most common handheld devices today, a Smartphone is usually

65

either a phone with added PDA capabilities or a PDA with added phone capabilities (as shown in Fig.12-2). Here's a list of some of the things Smartphones can do.

- Send and receive mobile phone calls—some Smartphones are also Wi-Fi capable
- Personal Information Management (PIM) including notes, calendar and to-do list
- Communication with laptop or desktop computers
- Data synchronization with applications like Microsoft Outlook and Apple's iCal calendar programs
- E-mail
- Instant messaging
- Applications such as word processing programs or video games
- Play audio and video files in some standard formats

Fig.12-2 Smartphone is a phone with added PDA capabilities

While most cell phones include some software, even the most basic models these days have address books or some sort of contact manager.

New Words

assistant [ə'sistənt] n.	助手，助理
built-in ['bilt'in] adj.	嵌入的，内置的
calendar ['kælində] n.	日历
stuck [stʌk] adj.	动不了的，被卡住的
tweak [twi:k] vt.	稍稍调整（机器、系统等）
handheld ['kænd,held] adj.	手持式的，便携式的，掌上的

Phrases and Expressions

tossed around	翻来覆去

Unit Ⅱ　Electronic Instruments and Products 电子仪器和设备

in a nutshell	简而言之
PDA (Personal Digital Assistants)	个人数字助理
cell phones	移动电话
adapt to	使适应于
Palm Pilot	掌上电脑
PIM (Personal Information Management)	（计算机的）个人信息管理程序
SYNC (SYNchronous Communication)	同步通信

Notes

1. In a nutshell, a Smartphone is a device that lets you make telephone calls, but also adds in features that, in the past, you would have found only on a personal digital assistant or a computer, such as the ability to send and receive E-mail and edit Office documents.

   译文：简而言之，智能手机是这样的一个设备，它能让你拨打电话，并且还增加了其他一些以前你只会在个人数字助理或计算机里发现的功能，例如，发送和接收 E-mail 或编辑 Office 文档。

   nutshell：n. 坚果壳，小容器，简明扼要。

   in a nutshell：简而言之，极其简括地，简单地说。

   E-mail：Electronic mail，n. 电子邮件。

   Office documents：Office 文档，办公文件。

2. Smartphones are run by using Operating Systems just like a computer does, some of the more popular operating systems or OS are the Android, iOS, Windows, Symbian and Blackberry OS.

   译文：智能手机的运行依靠的操作系统类似于计算机所用的，比较流行的操作系统有安卓、苹果、视窗、塞班和黑莓。

   Operating Systems：操作系统（OS）是用户和计算机之间的界面。一方面操作系统管理着所有的计算机系统资源，另一方面操作系统为用户提供了一个抽象概念上的计算机。在操作系统的帮助下，用户使用计算机时，避免了对计算机系统硬件的直接操作。

   Android：一种以 Linux 为基础的开放源代码操作系统，主要使用于便携设备中。在中国较多人使用"安卓"译名。

   iOS：苹果手机操作系统。

   Windows：微软公司生产的"视窗"操作系统。

   Symbian：n. 塞班（一种手机操作系统），是基于 EPOC32 的操作系统。

   Blackberry OS：黑莓 OS 是基于 Java 的 RIM 操作系统。

3. The Android OS was released in 2008 and it is considered an open source platform that was supported by Google.

   译文：安卓操作系统在 2008 年发布，它被认为是一个由谷歌支持的源程序开放的平台。

   Google：一家在全球享有盛誉的 Internet 搜索引擎。Google 是英文单词"Googol"按照通

67

常的英语拼法改写而来的。Googol 是一个大数的名称，它是 10 的 100 次方，表示 1 后面跟 100 个零。下面的一串就表示 Googol。看上去好像没什么了不起，是吗？但是它比宇宙中所有基本粒子的数量总和还要大。

4. The features it included were Maps, Calendar, Gmail and a fully functioning HTML web browser. The extremely popular Smartphone is the iPhone from Apple.

译文：它的功能包括了地图、日历、Gmail 和一个全功能的网页浏览器。最受欢迎的手机之一是苹果公司的 iPhone 手机。

Gmail：Google 的免费网络邮件服务。

HTML：Hyper Text Markup Language 的缩写，超文本标记语言。

Web：本意是蜘蛛网、网状物、网状组织。在这里是指一种超文本信息系统，Web 的一个主要的概念就是超文本链接。

iPhone：网络电话。iPhone 是结合照相手机、个人数字助理（PDA）、媒体播放器及无线通信设备的掌上设备，由苹果公司推出。

Apple：苹果公司（美国），是全球第一大手机生产商和主要的 PC 厂商，最知名的产品是其出品的 Apple II、Macintosh 电脑、iPod 音乐播放器、iTunes 商店、iMac 一体机、iPhone 手机和 iPad 平板电脑等。

Exercises

1. Write T (True) or F (False) beside the following statements about the text.

a. Cell phones were used for making calls.

b. PDAs gained wireless connectivity and were not able to send and receive E-mail.

c. Smartphones allow individual users to install, configure and run applications of their choosing.

d. Most standard cell-phone software offers a lot of choices for re-configuration.

e. If that phone were a Smartphone, you couldn't install any compatible calendar application you like.

2. Fill in the missing words according to the text.

a. A PDA could store your contact info and a _____ list, and could SYNC with your computer.

b. Smartphones may allow you to _____ and _____ Microsoft documents or at least view the files.

c. Smartphones are run by using _____ just like a computer does.

d. The iPhone was the first phone to have _____ that was designed for _____ input.

e. Since cell phones and PDAs are the most common handheld devices today, a Smartphone is usually either a phone with added _____ capabilities or a PDA with added _____ capabilities.

3. Translate the following paragraph into Chinese.

Since cell phones and PDAs are the most common handheld devices today, a Smartphone is usually either a phone with added PDA capabilities or a PDA with added phone capabilities. Here's a list of some of the things Smartphones can do.

- Send and receive mobile phone calls—some Smartphones are also Wi-Fi capable
- Personal Information Management (PIM) including notes, calendar and to-do list
- Communication with laptop or desktop computers
- Data synchronization with applications like Microsoft Outlook and Apple's iCal calendar programs
- E-mail
- Instant messaging
- Applications such as word processing programs or video games
- Play audio and video files in some standard formats

Chinese Translation of Texts（参考译文）

第 12 课　智能手机

你可能翻来覆去听到过"智能手机"这个术语很多次了。但如果你曾经想要确切地知道究竟什么是智能手机，好吧，不只你一个人想要知道。智能手机和普通手机相比到底有什么不同，是什么使它这么智能呢？

简而言之，智能手机是这样的一个设备，它能让你拨打电话，并且还增加了其他一些以前你只会在个人数字助理或计算机里发现的功能，例如，发送和接收 E-mail 或编辑 Office 文档。

但是，要真正理解什么是智能手机（或者不是），我们应该开始一堂历史课。起初，在开始的时候，有手机和个人数字助理（PDA）。手机除了用来拨打电话外没有别的功能，而 PDA，如掌上电脑，是作为个人的、便携式的助理器使用的。PDA 可以存储你的联系信息和待办事宜清单，并且可以与你的计算机同步。

PDA 获得了无线连接，并能够发送和接收电子邮件。与此同时，手机也获得了通信功能。PDA 中添加了更多类似手机的功能，而手机中添加了更多类似 PDA（甚至类似计算机）的功能。结果就出现了智能手机。

与许多传统手机不同，智能手机允许个人用户安装、配置和运行他们所选择的应用程序（见图 12-1）。智能手机具有可以使设备遵从你特定的做事方式的能力。最标准的手机软件只为重新配置手机提供有限的选择，迫使你去适应它的安装方式。在一个标准的手机里，不管你喜欢或者不喜欢内置的日历应用程序，除了做一些小调整，你都还是得接受。如果该手机是智能手机，你可以安装任何你喜欢的可以兼容的日历应用程序。例如，智能手机将使你能够做得更多。它可以让你创建和编辑微软文件或至少查看文件。它可以让你下载 Apps 等。

智能手机的运行依靠的操作系统类似于计算机所用的，比较流行的操作系统有安卓、苹果、视窗、塞班和黑莓。

安卓操作系统在 2008 年发布，它被认为是一个由谷歌支持的源程序开放的平台，它的功能包括了地图、日历、Gmail 和一个全功能的网页浏览器。最受欢迎的手机之一是苹果公司的 iPhone 手机。它在 2007 年第一次向全世界发布，售价高达 499 美元。它是第一个使用大触摸屏并设计为用手指直接输入的手机。

由于手机和掌上电脑是目前最常见的手持设备，智能手机通常既可以说是一个添加了掌上电脑功能的电话，又可以说是一个添加了电话功能的掌上电脑（见图 12-2）。下面是智能手机可以做的一些事情。

- 发送和接收移动电话的呼叫——一些智能手机还具有无线上网的功能。
- 个人信息管理（PIM）功能包括便条、日历和待办事宜清单。
- 与便携式计算机或台式计算机通信。
- 与应用程序进行数据同步，如 Microsoft Outlook 和苹果的 iCal 日历程序。
- 电子邮件。
- 即时消息。
- 应用程序，如 Word 处理程序或者视频游戏。
- 以一些标准格式播放音频和视频文件。

大多数手机都安装了许多软件，即使是最基本的手机也有通讯录和联系人管理器。

Reading Material

7. Digital Voltmeter

A voltmeter is used to measure the voltage in volts. A voltmeter is always connected across the two points of the device under test. Since voltmeters are always connected in parallel with the component or components under test, any current through the voltmeter will contribute to the overall current in the tested circuit, potentially affecting the voltage being measured.

An electronic digital voltmeter gives a reading on a numerical display (e.g. LED). It eliminates errors due to parallax that can arise in instruments requiring the position of a pointer on a scale to be estimated and also has a very high input resistance (e.g. 10 MΩ).

The simplified block diagram and the waveforms are shown in Fig.R7-1, which helps us to follow the action. The DC voltage to be measured is fed to one input of a voltage comparator. The other input of the comparator is supplied by a ramp generator which produces a repeating sawtooth waveform. The output from the comparator is "high" (1) until the ramp voltage equals the input voltage when it goes "low" (0).

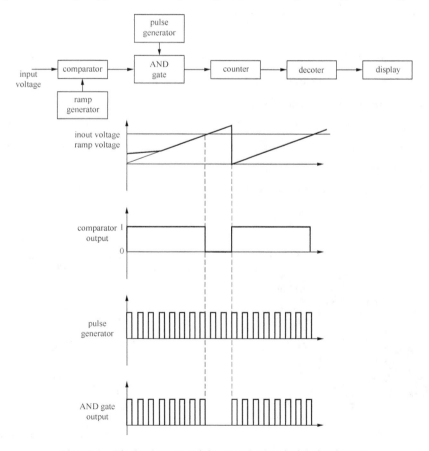

Fig.R7-1　Block Diagram and the Waveforms of Digital Voltmeter

The comparator output is applied to one input of an AND gate, the other input of the gate being fed by a generator. When both these inputs are "high", the gate opens and gives a "high" output, i.e. a pulse. The number of steady train of pulses from a pulse output pulses so obtained from the AND gate depends on the length of the comparator output pulse, i.e. on the time taken by the ramp voltage to reach the value of the input voltage. If the ramp is linear, this time is proportional to the input voltage.

The output pulses from the AND gate are recorded by a binary counter and then converted into decimal form by a decoder before being passed on to the display.

The whole process commences when the voltmeter is switched on and a pulse from a trigger circuit starts the ramp generator and sets the counter to zero. When the input voltage is of the order of millivolts, it is amplified before being measured. With some additional circuitry the voltmeter can be adapted for use as a multimeter to measure AC voltages, current and also resistance.

8. Touchscreen

A touchscreen is an intuitive computer input device that works by simply touching the display screen, either by a finger, or with a stylus, rather than typing on a keyboard or pointing with a mouse (Fig.R8-1).

The touchscreen interface—whereby users navigate a computer system by touching icons or links on the screen itself—is the most simple, intuitive, and easiest to learn of all PC input devices and is fast becoming the interface of choice for a wide variety of applications.

Public Information Systems (as Fig.R8-2): Information kiosks, tourism displays, and other electronic displays are used by many people that have little or no computing experience. The user–friendly touchscreen interface can be less intimidating and easier to use than other input devices, especially for novice users, making information accessible to the widest possible audience.

Fig.R8-1   Touchscreen

Fig.R8-2   Public Information Systems

Restaurant/POS Systems: Time is money, especially in a fast paced restaurant or retail environment. Because touchscreen systems are easy to use, overall training time for new employees can be reduced. And work can get done faster, because employees can simply touch the screen to perform tasks, rather than entering complex key strokes or commands.

Customer Self-Service: In today's fast pace world, waiting in line is one of the things that have yet to speed up. Self-service touchscreen terminals can be used to improve customer service at busy stores, fast service restaurants, transportation hubs, and more.

Control/Automation Systems: The touchscreen interface is useful in systems ranging from industrial process control to home automation. By integrating the input device with the display, valuable workspace can be saved. And with a graphical interface, operators can monitor and control complex operations in real-time by simply touching the screen.

Any touchscreen system comprises the following three basic components (as Fig.R8-3).

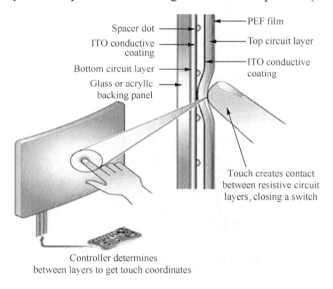

Fig.R8-3　Components of Touchscreen System

A touchscreen sensor panel, that sits above the display and which generates appropriate voltages according to where, precisely, it is touched.

A touchscreen controller, that processes the signals received from the sensor and translates there into touch event data which is passed to the PC's processor, usually via a serial or USB interface.

A software driver provides an interface to the PC's operating system and which translates the touch event data into mouse events, essentially enabling the sensor panel to "emulate" a mouse.

There are several types of touchscreens. Here we discuss the infrared touchscreens. Infrared touchscreens are based on light-beam interruption technology. Instead of placing a layer on the display surface, a frame surrounds it. The frame assembly is comprised of printed wiring boards on which the opto-electronics are mounted and is concealed behind an IR-transparent bezel.

The frame contains light sources—or light-emitting diodes—on one side, and light detectors-or photosensors—on the opposite side. The effect of this is to create an optical grid across the screen. When any object touches the screen, the invisible light beam is interrupted, causing a drop in the signal received by the photosensors. Based on which photosensors stop receiving the light signals, it is easy to isolate a screen coordinate.

Infrared touch systems are solid state technology and have no moving mechanical parts. As such, they have no physical sensor that can be abraded or worn out with heavy use over time. Furthermore, since they do not require an overlay—which can be broken—they are less vulnerable to vandalism and also extremely tolerant of shock and vibration.

9. High Definition TV (HDTV)

HDTV or High Definition TV (as shown in Fig.R9-1) refers to those amazingly realistic audio and video signals received via cable or satellite. Your decoder receives these signals and converts them to high quality audio/video. HDTV implies a larger aspect ratio—wide screen TV in your home theater system, the way movies were always meant to be viewed. HDTV resolution is far superior compared to traditional TV, which is why it is preferred.

Fig.R9-1　High Definition TV

A conventional, analog TV uses a cathode ray tube to deliver images to you, which limits the quality of the image. The screen resolution of an analog TV is about 512×400 pixels. HDTV uses a digital display, like your computer monitor, and the screen resolution is at least 1280×720 pixels, which is comparable to a high-end computer display. A higher screen resolution means a crisper, clearer picture. In addition to dramatically improved picture quality, HDTV also offers a wider format. This makes an HDTV image more like a movie-screen image. The width-to-height ratio—called the aspect ratio—of HDTV is 16:9. Analog TV has an aspect ratio of only 4:3.

The difference in aspect ratio is most noticeable when watching theatrical movies on TV. For analog TV, the movie must be cut down in a process called "pan and scan", in which a part of every scene is deleted to fit the lower aspect ratio. The only way to see the entire movie scene on an analog TV is to "letterbox" the movie. In letterboxing, the full movie is shown in the middle of the screen with black bars at the top and bottom. HDTV eliminates letterboxing and allows you to see the complete movie on the whole TV screen.

The Federal Communications Commission (FCC) agrees with TV networks and manufacturers that digital television, including HDTV, should be the new standard for broadcasting. As of May 1999, the FCC requires the top TV networks to broadcast a digital signal in the 10 biggest markets, which represent 30 percent of TV households in the US. The networks plan to expand digital coverage and phase out analog TV broadcasts entirely by the end of 2006.

10. Digital Audio

Digital audio is the representation of audio information in digital (discrete level) formats. The use of digital audio allows for simple transmission, storage, and processing of audio signals.

Audio Digitization—Audio digitization is the conversion of analog audio sounds into digital form. To convert an analog audio signal to digital form, the analog signal is digitized by using an analog-to-digital (pronounced A to D) converter. The A/D converter periodically senses (samples) the level of the analog signal and creates a binary number or series of digital pulses that represent the level of the signal (as shown in Fig.R10-1). The typical sampling rate for the conversion of analog audio ranges from 8,000 samples per second (for telephone quality) to 44,000 samples per second (for music quality).

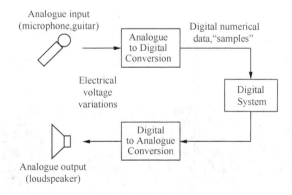

Fig.R10-1   A/D Converter and D/A Converter

Audio Compression—Audio compression is the analysis and processing of digital sound to a form that reduces the space required for transmission or storage. Audio compression coders and decoders (codecs) analyze digital audio signals to remove signal redundancies and sounds that cannot be heard by humans.

Digital audio data is random in nature unlike digital video which has repetitive information that occurs on adjacent image frames. This means that audio signals do not have a high amount of redundancy, making traditional data compression and prediction processes ineffective at compressing digital audio. It is possible to highly compress digital audio by removing sounds that can be heard or perceived by listeners through the process of perceptual coding.

Audio Coding—The type of coder (type of analysis and compression) can dramatically vary and different types of coders may perform better for different types of audio sounds (e.g. speech audio as compared to music).

The MPEG system allows for the use different types of audio coders. The type of coder that is selected can vary based on the application (such as playing music or speech) and the type of device the audio is being played through (such as a television or a battery operated portable media player). The MPEG speech coders range from low complexity (layer 1) to high complexity (layer 3). A new version of audio coder has been created (advanced audio codec-AAC) that offers better audio quality at lower bit rates. The AAC coder also has several variations that are used in different types of applications (e.g. broadcast radio -vs.- real time telephony).

Multichannel Audio—Multichannel audio is the use of multiple sound channels to produce an enhanced listening experience. Examples of multichannel audio include stereo, surround sound and Low Frequency Enhancement (LFE). The MPEG system was designed to allow the combination of multiple audio channels (such as stereo).

Many types of audio applications can benefit from the high quality, power efficiency, space savings and ease of use in TI's all-digital audio solution. These include consumer entertainment such as HTIB, DVD receivers and mini/micro systems, PC gaming and entertainment applications, business applications, such as speakerphones and multimedia conferencing systems, and digital headsets for all application areas. In addition, TI's all-digital audio solution will enhance automotive entertainment, reducing space and heat dissipation for manufacturers while allowing consumers to enjoy the experience of high-quality multichannel digital audio in their vehicles (as shown in Fig.R10-2).

Fig.R10-2　Digital Audio System

11. Home Theater System

A home theater consists of audio and video equipments that transform your home into a movie theater. Your home theater system could comprise of a large screen TV, a DVD player an AV receiver or stereo and speakers (as shown in Fig.R11-1). Anything else is optional and will enhance your enjoyment. Home theater design is fundamentally based on how easy it is to use and set up, easy maintenance, good performance and looks, of course. Your home theater system can be located anywhere in your home convenient to you.

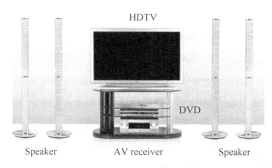

Fig.R11-1 Home Theater System

Home theater systems are designed to provide the real theater like environment. Besides the visual effects, sound is very crucial to creating the real theater like environment. You must have watched movies in a theater. One thing that always captivates the audience is the sound—the Dolby Surround sound, which makes you feel like you are in middle of the action.

So what is Dolby surround sound? The Dolby Surround process involves encoding four channels of information into a two-channel signal. A decoder then decodes the signal and sends it to appropriate destination—Left, Right, Rear, and Phantom Center. In Dolby Surround, the main sounds come from the left and right channels, the vocal or dialog come from the center Phantom channel, and finally the ambience effect is generated from the rear channel.

Dolby surround system was the first to introduce surround sound in home theater system. It evolved into Dolby Surround Pro Logic, Dolby Digital EX, and THX Surround EX. Presently the current standard in surround sound is Dolby Digital 5.1 channel. Dolby Digital is widely used in laser discs, satellite programming, and DVDs. To get Dolby Digital surround sound you should have a discrete Dolby surround sound decoder in the DVD player and Dolby surround sound preamplifier or Dolby digital surround sound receiver.

If you have always coveted a particular type of home theater system, you can take those home theater pictures when you go shopping for your system. It will also help if you have a professional helping you with the home theater design. If you enjoy do-it-yourself projects, you can check around and get information on different types of systems. Home theater magazines and home theater system reviews online can help a great deal.

After making your plan for how exactly you would like to set up your home theater, next comes creating the atmosphere with home theater furniture, setting up speakers, wiring, lighting, storage cabinets, etc. Seating is an important aspect of your home theaters. You can have recliners, couches, or just regular chairs. Both type of seating and locating them is an important decision. Depending on how much space you have, you can decide how many seats to install, as you will have to maintain the right angle and distance from the screen.

12. Set-Top Box (STB)

A Set-Top Box (STB) is an information appliance device that generally contains a tuner and connects to a television set and an external source of signal, turning the source signal into content in a form that can then be displayed on the television screen or other display device. Set-top boxes can also enhance source signal quality. They are used in cable television (as Fig.R12-1) and satellite television systems, as well as other uses.

Fig.R12-1　A Set-Top Box (STB)

Set-top boxes were also made to enable closed captioning on older sets in North America, before this became a mandated inclusion in new TV sets. Some have also been produced to mute the audio (or replace it with noise) when profanity is detected in the captioning, where the offensive word is also blocked. Some also include a V-chip that allows only programs of some television content ratings. A function that limits children's time watching TV or playing video games may also be built in, though some of these work on main electricity rather than the video signal.

Hybrid set-top boxes, such as those used for Smart TV programming, enable viewers to access multiple TV delivery methods (including terrestrial, cable, internet, and satellite). Like IPTV boxes, they include video on demand, time-shifting TV, Internet applications, video telephony, surveillance, gaming, shopping, TV-centricelectronic program guides, and e-government. By integrating varying delivery streams, hybrids (sometimes known as "TV-centric") enable pay-TV operators more flexible application deployment, which decreases the cost of launching new services, increases speed to market, and limits disruption for consumers.

As examples, Hybrid Broadcast Broadband TV (HBBTV) set-top boxes allow traditional TV broadcasts, whether from terrestrial (DTT), satellite, or cable providers, to be brought together with video delivered over the Internet and personal multimedia content. Advanced Digital Broadcast (ADB) launched its first hybrid DTT/IPTV set-top box in 2005, which provided Telefónica with the digital TV platform for its Movistar Imagenio service by the end of that year.

UK based Inview Technology has over 8M STBs deployed in the UK for Teletext and an original push VOD service for Top Up TV.

In IPTV networks, the set-top box is a small computer providing two-way communications on an IP network and decoding the video streaming media. IP set-top boxes have a built-in home network interface that can be Ethernet or one of the existing wire home networking technologies such as HomePNA or the ITU-T G.hn standard, which provides a way to create a high-speed (up to 1Gbit/s) local area network using existing home wiring (power lines, phone lines, and coaxial cables).

In the US and Europe, telephone companies use IPTV (often on ADSL or optical fiber networks) as a means to compete with traditional local cable television monopolies.

Unit III

Communicated Technology

通信技术

本章重点介绍了通信技术的应用，主要包括光纤通信、卫星通信、Wi-Fi、全球定位系统、无线传感网、4G 网络等。通过本章，学生可以学到关于通信技术应用的较为系统的英文表达，为在国际性很强的通信技术领域职业工作环境中应用英语打下基础。

Lesson 13 Optical Fiber Communications

Communication may be broadly defined as the transfer of information from one point to another. When the information is to be conveyed over any distance, a communication system is usually required. Within a communication system, the information transfer is frequently achieved by superimposing or modulating the information on to an electromagnetic wave which acts as a carrier for the information signal. This modulated carrier is then transmitted to the required destination where it is received and the original information signal is obtained by demodulation. Sophisticated techniques have been developed for this process by using electromagnetic carrier waves operating at radio frequencies as well as microwave and millimeter wave frequencies. However, "communication" may also be achieved by using an electromagnetic carrier which is selected from the optical range of frequencies.

Typical optical fiber communications system is shown in Fig.13-1. In this case, the information source provides an electrical signal to a transmitter comprising an electrical stage which drives an optical source to give modulation of the lightwave carrier. The optical source which provides the electrical-optical conversion may be either a semiconductor laser or Light Emitting Diode (LED). The transmission medium consists of an optical fiber cable and the receiver consists of an optical detector which drives a further electrical stage and hence provides demodulation of the optical carrier.

Photodiodes (p-n, p-i-n or avalanche) and, in some instances, phototransistor and photoconductors are utilized for the detection of the optical signal and the optical-electrical conversion. Thus there is a requirement for electrical interfacing at either end of the optical link and at present the signal processing is usually performed electrically.

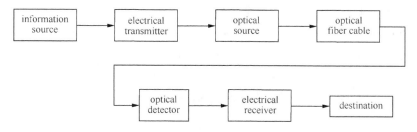

Fig. 13-1　Optical Fiber Communications System

The optical carrier may be modulated by using either an analog or digital information signal. Analog modulation involves the variation of the light emitted from the optical source in a continuous manner. With digital modulation, however, discrete changes in the light intensity are obtained (i.e. on-off pulses). Although often simpler to implement, analog modulation with an optical fiber communication system is less efficient, requiring a far higher signal to noise ratio at the receiver than digital modulation. Also, the linearity needed for analog modulation is not always provided by semiconductor optical source, especially at high modulation frequencies. For these reasons, analog optical fiber communication link are generally limited to shorter distances and lower bandwidths than digital links.

Initially, the input digital signal from the information source is suitably encoded for optical transmission. The laser drive circuit directly modulates the intensity of the semiconductor laser with the encoded digital signal. Hence a digital optical signal is launched into the optical fiber cable. The Avalanche Photodiode Detector (APD) is followed by a front-end amplifier and equalizer or filter to provide gain as well as linear signal processing and noise bandwidth reduction. Finally, the signal obtained is decoded to give the original digital information.

New Words

convey [kən'vei] vt.	搬运，传达，转让
transfer ['trænsfə:] n.	移动，传递，转移
superimpose [ˌsu:pərɪm'pəʊz] v.	添加，双重
microwave ['maikrəweiv] n.	微波
conversion [kən'və:ʃn] n.	变换，转化
semiconductor [ˌsemikən'dʌktə] n.	半导体
diode ['daiəud] n.	二极管
photodiode [ˌfəutəu'daiəud] n.	光敏二极管，光电二极管

phototransistor [ˌfəutəutræn'zistə] n.	光电晶体管，光敏晶体管
photoconductor [ˌfəutəukən'dʌktə] n.	【物】光电导体，光电导元件
discrete [di'skri:t] adj.	不连续的，离散的
implement ['implimənt, 'impliment] n. & vt.	工具，器具；贯彻，实现，执行
encode [in'kəud] vt.	编码
equalizer ['i:kwəlaizə] n.	均衡器，平衡装置
decode [ˌdi:'kəud] vt.	解码，译解

Phrases and Expressions

optical fiber	光纤
electromagnetic wave	电磁波
APD (Avalanche PhotoDiode)	雪崩发光二极管
front-end amplifier	前置放大器

Notes

1. …by superimposing or modulating the information on to an electromagnetic wave which acts as a carrier for the information signal.

   （1）介词 by + 动名词短语充当状语。（2）which 引导定语从句，修饰 wave。

2. This modulated carrier is then transmitted to the required destination where it is received and the original information signal is obtained by demodulation.

   译文：这一经过调制的载波随后被传送到要求到达的目的地，在那里被接收，并且通过解调还原成原始信息。

   过去分词短语 required 充当定语，where it is …by demodulation 是定语从句，两者均修饰 destination。

3. … by using electromagnetic carrier waves operating at radio frequencies as well as microwave and millimeter wave frequencies.

   （1）介词 by + 动名词短语充当状语。（2）as well as:"以及"，相当于 and，常用于句中，下文中仍有这种用法。再如，He can speak English as well as Chinese.

   请比较：He can speak English, and he can speak Chinese as well.（as well 用于句末，相当于 too）。

4. However, "communication" may also be achieved by using an electromagnetic carrier which is selected from the optical range of frequencies.

   译文：在通信中，也可选择光波的频率作为载波频率。

   which 引导定语从句，修饰 electromagnetic carrier。

5. In this case the information source provides an electrical signal to a transmitter comprising an electrical stage which drives an optical source to give modulation of the lightwave carrier.

译文：信源提供电信号给发射机，发射机组成一个电子平台来驱动光源以完成对光载频的调制。

（1）in this case：在这种场合下、既然这样。（2）which 引导定语从句。

6. …to provide gain as well as linear signal processing and noise bandwidth reduction.

此为不定式短语，充当目的状语。

Exercises

1. Write T (True) or F (False) beside the following statements about the text.

a. This modulated carrier is then transmitted to the required destination where it is received and the original information signal is obtained by modulation.

b. Sophisticated techniques have been developed for this process by using magnetic carrier waves operating at radio frequencies as well as microwave and millimeter wave frequencies.

c. The information source provides an electrical signal to a transmitter comprising an electrical stage which drives an optical source to give modulation of the lightwave carrier.

d. The transmission medium consists of an optical fiber cable and the receiver.

e. The optical carrier is modulated by using neither an analog nor digital information signal.

f. With digital modulation, discrete changes in the light intensity are not obtained.

g. Analog modulation with an optical fiber communication system is more efficient, requiring a far lower signal to noise ratio at the receiver than digital modulation.

h. The input digital signal from the information source is not suitably encoded for optical transmission.

2. Fill in the missing words according to the text.

a. Within a communication system the information transfer is _____ by superimposing or modulating the information on to an electromagnetic wave which acts as a carrier.

b. Photodiodes (p-n, p-i-n or avalanche) and _____ and _____ are utilized for the detection of the optical signal and the optical-electrical conversion.

c. The laser drive circuit directly modulates the intensity of the semiconductor laser with the encoded digital signal. Hence a _____ is launched into the optical fiber cable.

d. There is a requirement for electrical interfacing at either end of the optical link and at present the _____ is usually performed electrically.

3. Translate the following paragraph into Chinese.

Communication may be broadly defined as the transfer of information from one point to another. When the information is to be conveyed over any distance, a communication system is usually required. Within a communication system, the information transfer is frequently achieved by superimposing or modulating the information on to an electromagnetic wave which acts as a carrier for the information signal. This modulated carrier is then transmitted to the required destination where it is received and the original information signal is obtained by demodulation. Sophisticated techniques have been developed for this process by using electromagnetic carrier waves operating at radio frequencies as well as microwave and millimeter wave frequencies. However, "communication" may also be achieved by using an electromagnetic carrier which is selected from the optical range of frequencies.

Chinese Translation of Texts（参考译文）

第 13 课　光纤通信

广义地讲，把信息从一点传送到另一点就称为通信。当信息跨越一段距离被传送时，就需要一个通信系统。在通信系统中，信息传送是通过把信息叠加在电磁波上或对电磁波进行调制来实现的，电磁波起着装载信号的作用。这一经过调制的载波随后被传送到要求到达的目的地，在那里被接收，并且通过解调还原成原始信息。在运用电磁载波的领域，高新技术得到进一步的发展，如射频、微波以及毫米波的频率都被用来作为载波频率。在通信中，也可选择光波的频率作为载波频率。

典型的光纤通信系统如图 13-1 所示。信源提供电信号给发射机，发射机组成一个电子平台来驱动光源以完成对光载频的调制。光源是由发光二极管或半导体激光管构成的，它可完成光电转换。传输媒介由光纤（缆）组成。光接收机包括一个含光检测器的电路驱动平台，用以完成对已调光载波的解调。用于检测光信号和进行光电转换的器件有光电二极管（p-n、p-i-n 或雪崩管），在某些情况下，还有发光三极管及光敏电阻等。因此，在光系统链路的两端都要求有电接口，并且在现阶段，信号处理通常是通过电路实现的。

模拟或数字的信号均可用来调制光载波。模拟调制是指从光源处发射的连续光强度的变化，而数字调制则不然，它是通过光强度离散的变化（如有无光脉冲）来实现的。模拟调制的实施相对简单，但在光系统调制中调制效率较低，而且与数字调制相比，需要高得多的信噪比。模拟调制所必需的线性度并不总是半导体光源提供，尤其是在高频调制中。基于上述原因，与数字光系统相比较，模拟通信链路通常被限制使用在较短的通信距离和较窄的带宽上。

首先，信源的数字信号被适当地编码以进行光传输。激光器的驱动电路通过这些已编码的数字信号来直接调制激光器的发光强度，然后数字光信号被注入光纤。在接收端，信号通过雪崩发光二极管（APD）后进入前置放大器和均衡器或滤波器，放大器用来提供增益、滤波器用来对信号进行线性处理和减少噪声带宽。最后，对获得的信号进行解调得到原始信号。

Lesson 14 Satellite Communications

Satellite communication has become a part of everyday life in the late 1980s. An international telephone call is made as easily as a local call to a friend who lives down the block. We also see international events, such as an election in England and a tennis match in France, with the same regularity as local political and sporting events. In this case, a television news program brings the sights and sounds of the world into our homes each night.

This capability to exchange information on a global basis, be it a telephone call or a news story, is made possible through a powerful communications tool—the satellite. For those of us who grew up at a time when the space age was not a part of everyday life, satellite-based communication is the culmination of a dream that stretches back to an era when the term satellite was only an idea conceived by a few inspired individuals. These pioneers included authors such as Arthur C.Clarke, who fostered the idea of a worldwide satellite system in 1945. This idea has subsequently blossomed into a sophisticated satellite network that spans the globe.

The latter type of satellite system would have entailed the development of a very complex and cumbersome earth and space-based network. Fortunately though, this problem was eliminated in 1963 and 1964 through the launching of the Syncom satellite. Rather than circling the Earth at a rapid rate of speed, the spacecraft appeared to be stationary or fixed in the sky. Today's communications satellites, for the most part, have followed suit and are now placed in what are called geostationary orbital positions or "slots".

Simply stated, a satellite in a geostationary orbital position appears to be fixed over one portion of the Earth. At an altitude of 22,300 miles above the equator, a satellite travels at the same speed at which the rotates, and its motion is synchronized with the Earth's rotation. Even though the satellite is moving at an enormous rate of speed, it is stationary in the sky in relation to an observer on the earth.

The primary value of a satellite in a geostationary orbit is its ability to communicate with ground stations in its coverage area 24 hours a day. This orbital slot also simplifies the establishment of the communications link between a station and the satellite. Once the station's antenna is properly aligned, only minor adjustments may have to be made in the antenna's position over a period of time. The antenna is repositioned to a significant degree only when the station establishes contact with a satellite in a different slot. Prior to this era, a ground station's antenna had to physically track a satellite as it moved across the sky.

Based on these principles, three satellites placed in equidistant positions around the earth can create a worldwide communications system in that almost every point on the earth can be reached by satellite (as shown in Fig.14-1).

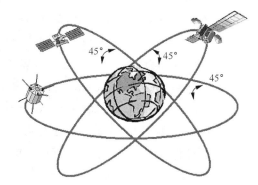

Fig.14-1 Satellite Communication

This concept was the basis of Arthur Clarke's original vision of a globe-spanning communications network.

New Words

block [blɔk] n.	街区
global ['gləubl] adj.	球形的，全球的，全世界的
culmination [ˌkʌlmi'neiʃn] n.	顶点
era ['iərə] n.	时代，纪元，时期
span [spæn] v. & n.	横越，跨度；跨距，范围
cumbersome ['kʌmbəsəm] adj.	讨厌的，麻烦的，笨重的
syncom ['siŋkɔm] n.	（美国的）同步通信卫星
geostationary [ˌdʒi:əu'steiʃnri] n.	与地球的相对位置不变的，相对地静止的
altitude ['æltitju:d] n.	（尤指海拔）高度
equator [i'kweitə] n.	赤道
enormous [i'nɔ:məs] adj.	巨大的
equidistant [ˌi:kwi'distənt] adj.	距离相等的，等距的

Phrases and Expressions

tennis match	网球比赛
prior to	在前，居先
globe-spanning communications network	全球通信网络

Notes

1. An international telephone call is made as easily as a local call to a friend who lives down the block.

   译文：打国际电话就像给住在同一街区的朋友打本地电话一样简单。

   as easily as: 像……一样（容易），easily 修饰动词 made，充当状语。请比较：Our classroom is as big and bright as theirs.（形容词 big and bright 充当表语）。

2. This capability to exchange information on a global basis, be it a telephone call or a news story, is made possible through a powerful communications tool—the satellite.

   译文：这种进行全球性信息交流的能力，无论是电话还是新闻转播，都是通过一个强有力的通信工具——卫星，才得以实现的。

3. …that stretches back to an era when the term satellite was only an idea conceived by a few inspired individuals.

(1) that 引导定语从句，修饰 dream。(2) to stretch back: 回顾、追溯，相当于 to recall back。(3) an era 指全球卫星通信系统仍是人们头脑中灵感的想象的年代，其后有 when 引导的定语从句修饰。

4. These pioneers included authors such as Arthur C.Clarke, who fostered the idea of a worldwide satellite system in 1945.

译文：这些先驱者中包括亚瑟·克拉克，他在1945年就产生了全球卫星系统的想法。

句中，who 引导非限制性定语从句，修饰 Arthur C.Clarke。

5. The latter type of satellite system would have entailed the development of a very complex and cumbersome earth and space-based network.

译文：亚瑟·克拉克设想的卫星通信系统必须开发复杂和烦琐的地面和空间网络。

latter: 形容词，后面的，（两者中）后者的，近来的，指顺序的先后，而 later 指时间的先后。

6. Rather than circling the earth at a rapid rate of speed, the spacecraft appeared to be stationary or fixed in the sky.

译文：现在的卫星不是高速地围绕地球运行，而是静止或固定在空中。

(1) rather than…: 而不是……，置于句首可起强调作用。(2) to appear to do sth.: 似乎/碰巧做某事，后文中仍有此用法。

7. Simply stated, a satellite in a geostationary orbital position appears to be fixed over one portion of the earth.

译文：简单地讲，一颗处于与地球旋转同步的轨道位置上的卫星对于地球的部分地区来说似乎是静止不动的。

simply stated: 简单地说，相当于 simply speaking，请比较: generally speaking（总的来说）。

8. Prior to this era, a ground station's antenna had to physically track a satellite as it moved across the sky.

译文：在这以前，地面站的天线必须物理跟踪在天空中运动的卫星。

(1) prior to this era: 在此之前、在这个年代之前，相当于 before this age。(2) as 引导时间状语从句。

Exercises

1. Write T (True) or F (False) beside the following statements about the text.

a. Satellite communication has become a part of everyday life in the late 1960s.

b. An international telephone call is still not easy as a local call.

c. We can see international events in our homes each night by the television news program.

d. Satellite communication brings the sights and sounds of the world via a television set.

e. Now satellite-based communication only is a dream by a few inspired individuals.

f. In l945, Arthur C.Clarke had developed a worldwide satellite system.

g. Today's communications satellites, for the most part, circling the earth at a rapid rate of speed.

h. The satellite is stationary in the sky in relation to an observer on the earth.

2. Match the following terms to appropriate definition or expression.

    a. satellite   1. create a world-wide communications

    b. "slots"   2. geostationary orbital positions

    c. three satellites   3. a powerful communications tool

3. Fill in the missing words according to the text.

    a. Fortunately though, _____ problem was eliminated in 1963 and 1964 through the launching of the Syncom satellite.

    b. Based on _____ principles, three satellites placed in equidistant positions around the earth can create a _____ system in that almost every point on the earth can be reached by satellite.

    c. This orbital slot also simplifies the establishment of the communications link between _____ and _____.

    d. Prior to this era, a ground station's _____ had to physically track a satellite as it moved across the _____.

4. Translate the following paragraph into Chinese.

The latter type of satellite system would have entailed the development of a very complex and cumbersome earth and space-based network. Fortunately though, this problem was eliminated in 1963 and 1964 through the launching of the Syncom satellite. Rather than circling the Earth at a rapid rate of speed, the spacecraft appeared to be stationary or fixed in the sky. Today's communications satellites, for the most part, have followed suit and are now placed in what are called geostationary orbital positions or "slots".

Chinese Translation of Texts（参考译文）

第 14 课 卫 星 通 信

    20 世纪 80 年代后期，卫星通信逐渐成为日常生活的一部分，打国际电话就像给住在同一街区的朋友打本地电话一样简单。同样，我们获悉国际大事，像英国的竞选、法国的网球赛就像得知本地政治、体育新闻一样平常。此时电视新闻节目每天晚上都将全世界的声音和画面带进我们的家中。

    这种进行全球性信息交流的能力，无论是电话还是新闻转播，都是通过一个强有力的通信工具——卫星，才得以实现的。对于我们中那些并非生长在太空时代的人们来讲，卫星通信是人们长期以来一种梦想的顶点，这个梦想可以一直追溯到卫星这个词只是几个天才头脑中灵感的想象那个年代。这些先驱者中包括亚瑟·克拉克，他在 1945 年就产生了全球卫星系统的想法，这个想法随后发展成为一个遍布全球的复杂卫星网络。

亚瑟·克拉克设想的卫星通信系统必须开发复杂和烦琐的地面和空间网络。幸运的是，这个问题随着 1963 年和 1964 年同步卫星的发射成功而消除。现在的卫星不是高速地围绕地球运行，而是静止或固定在空中，目前大部分通信卫星都被定位在相对地球静止的轨道或被称为"槽"的位置上。

简单地讲，一颗处于与地球旋转同步的轨道位置上的卫星对于地球的部分地区来说似乎是静止不动的，在赤道上空 22300 mile（即 35881 km）的高度上，卫星与地球以同样的角速度运行，即它的运转与地球的自转是同步的。尽管卫星是以很高的速度运行的，但对于一个地球上的观察者而言，它总是停留在天空中的同一个位置上。

位于一个对地相对静止的轨道上的卫星的主要作用在于它可一天 24 小时地与它覆盖的地面站保持联系。这个轨道槽的位置同时使建立卫星与地面站之间的通信链路更加简单化。当一个地面站的天线处在适当的位置时，在一段很长时间内天线的位置只需做微小的调整。只有当一个地面站要与另外一个轨道槽中的卫星建立联系时，天线的位置才需做显著的调整。在这以前，地面站的天线必须物理跟踪天空中运动的卫星。

基于这些原理，围绕地球等距离位置上放置三颗卫星就可建立一个全球的通信系统，以使地球上的每一个点都能与卫星相连（见图 14-1）。这个概念也是以亚瑟·克拉克的最早有关全球通信网络的蓝图为基础的。

Lesson 15 Wireless Fidelity

Wireless Fidelity, popularly known as Wi-Fi, developed on IEEE 802.11 standards, is a widely used technology advancement in wireless communication. As the name indicates, Wi-Fi provides wireless access to applications and data across a radio network. Wi-Fi sets up numerous ways to build up a connection between the transmitter and the receiver such as DSSS, FHSS, IR and OFDM.

Wi-Fi provides its users with the liberty of connecting to the Internet from any place such as their home, office or a public place without the hassles of plugging in the wires. Wi-Fi is quicker than the conventional modem for accessing information over a large network. With the help of different amplifiers, the users can easily change their location without disruption in their network access. Wi-Fi devices are compliant with each other to grant efficient access of information to the user. Wi-Fi location where the users can connect to the wireless network is called a Wi-Fi hotspot. Accessing a wireless network through a hotspot in some cases is cost-free while in some it may carry additional charges. Many standard Wi-Fi devices such as PCI, miniPCI, USB, Cardbus and PC card, ExpressCard make the Wi-Fi experience convenient and pleasurable for the users.

Wi-Fi uses radio networks to transmit data between its nodes. Such networks are made up of cells that provide coverage across the network. The more the number of cells, the greater and stronger is the coverage on the radio network. The most popular Wi-Fi technology such as 802.11b operates on the range of 2.40 GHz up to 2.4835 GHz band. This provides a comprehensive platform for operating Bluetooth strategy, cellular phones, and other scientific equipments. While 802.11a technology has the range of 5.725 GHz to 5.850 GHz and provides up to 54 Mbit/s in speed. 802.11g technology is even better as it covers three non-overlapping channels and allows PBCC.

Device that can use Wi-Fi (such as a personal computer, video game console, Smartphone, tablet, or digital audio player) can connect to a network resource such as the Internet via a wireless network access point. Wi-Fi networks have limited range. A typical Wi-Fi home router using 802.11b or 802.11g might have a range of 45 m (150 ft) indoors and 90 m (300 ft) outdoors. Range also varies, as Wi-Fi is no exception to the physics of radio wave propagation, with frequency band. Wi-Fi in the 2.4 GHz frequency block has better range than Wi-Fi in the 5 GHz frequency block, and less range than the oldest Wi-Fi (and pre-Wi-Fi) 900 MHz block.

Hotspot coverage can comprise an area as small as a single room with walls that block radio waves or as large as many square miles—this is achieved by using multiple overlapping access points (as shown in Fig.15-1).

Unit Ⅲ   Communicated Technology 通信技术

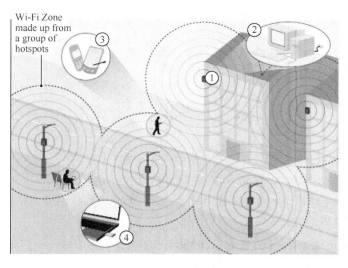

Fig.15-1   Wi-Fi Zone

New Words

fidelity [fi'deləti] n.	逼真，保真度
access ['ækses] vt.	使用，接近，获取
hassle ['hæsl] n.	麻烦事
plug [plʌg] v.	插上插头
disruption [dis'rʌpʃn] n.	破裂，毁坏，中断
console [kən'səul] n.	控制台，操纵台
propagation [ˌprɔpə'geiʃn] n.	传播，传输
comprise [kəm'praiz] vt.	包含，包括，由……组成
overlapping [ˌəuvə'læpiŋ] adj.	重叠的
router ['ru:tə] n.	路由器

Phrases and Expressions

DSSS (Direct Sequence Spread Spectrum)	直接序列扩频
OFDM (Orthogonal Frequency Division Multiplexing)	正交频分复用
FHSS (Frequency-Hopping Spread Spectrum)	跳频扩频
PCI（Peripheral Component Interconnect）	外设部件互连标准
SSID (Service Set IDentifier)	服务集标识符
USB (Universal Serial Bus)	通用串行总线
CardBus	具有总线控制功能的 32 位 PC 卡
PBCC (Packet Broadcast Control Channel)	分组广播控制信道

mini PCI	内置型无线网卡
Express Card	扩展接口
Mbit/s (Megabits per second)	兆比特/秒

Notes

1. Wireless Fidelity, popularly known as Wi-Fi, developed on IEEE 802.11 standards, is widely used technology advancement in wireless communication.

   译文：无线相容性认证，通常称为 Wi-Fi，它由 IEEE 802.11 标准发展而来，是在无线通信中被广泛应用的先进技术。

   IEEE：Institute Of Electrical and Electronics Engineers. 美国电气与电子工程师协会。

   IEEE 802.11 standards：美国电气与电子工程师协会标准。

2. Wi-Fi sets up numerous ways to build up a connection between the transmitter and the receiver such as DSSS, FHSS, IR – Infrared and OFDM.

   译文：Wi-Fi 在发射端和接收端之间设置了许多方法来建立连接，如扩频、跳频、红外的连接和正交频分复用。

   IR-Infrared：IR 是红外线的意思，常用说法 IR 接口，也就是红外线接口。

3. Many standard Wi-Fi devices such as PCI, miniPCI, USB, Cardbus and PC card, ExpressCard make the Wi-Fi experience convenient and pleasurable for the users.

   译文：许多标准的 Wi-Fi 设备，如 PCI、miniPCI、USB、Cardbus 及 PC 卡、ExpressCard，使得用户的 Wi-Fi 体验方便而愉快。

   PCI：Peripheral Component Interconnect，一种由英特尔（Intel）公司 1991 年推出的用于定义局部总线的标准。

   PC card：PC 卡，一种用以给计算机及其他通信和电子设备添加诸如存储器、大容量存储器、网络及无线通信等扩展装置的技术。

   miniPCI：miniPCI 插槽也同样是在 PCI 的基础上发展起来的，miniPCI 的定义与 PCI 基本上一致，只是在外形上进行了微缩。

   Express Card：由 PCMCIA 联盟推出的新规格，其优点是体积更小，传输速度更快，更适合移动系统。

   USB：通用串行总线（Universal Serial Bus），是一新型界面规格，支持主系统与不同外设间的数据传输。USB 允许外设在开机状态下插拔使用，USB 具有易于使用、高带宽、可接多达 127 个外设、数据传输稳定、支持即时声音播放及影像压缩等特点。目前在国内市场可以见到的 USB 设备主要有扫描仪、数码相机、打印机、集线器和外置存储设备等。

   CardBus：PCMCIA 推出的下一代高性能 32 位总线主控接口。它使现在只在桌面和较大的系统上才拥有的高级功能可以移入 CardBus 卡，从而可以用在移动环境下。

4. While 802.11a technology has the range of 5.725 GHz to 5.850 GHz and provides up to 54 Mbit/s in speed. 802.11g technology is even better as it covers three non-overlapping channels and

allows PBCC.

译文：虽然 802.11a 技术有着 5.725 GHz 到 5.850 GHz 的范围，并可提供高达 54 Mbit/s（兆比特/秒）的速度，但 802.11g 技术甚至更好，它包括 3 个非重叠信道，并允许分组广播控制信道。

802.11a technology：IEEE 于 1997 年公告的无线区域网络标准，适用于有线站台与无线用户或无线用户之间的沟通联结。802.11a 是 802.11 的衍生版，于 5.8 GHz 频段提供了最高 54 Mbit/s 的速率规格，并运用正交频分复用编码机制以取代 802.11 的 FHSS 或 DSSS。802.11g 在 2.4 GHz 频段上提供高于 20 Mbit/s 的速率规格。

5. A typical Wi-Fi home router using 802.11b or 802.11g might have a range of 45 m (150 ft) indoors and 90 m (300 ft) outdoors.

译文：一个使用 802.11b 或 802.11g 的典型 Wi-Fi 家庭路由器可能有 45 米（150 英尺）的室内范围和 90 米（300 英尺）的室外范围。

ft：英尺的意思，feet 复数，foot 单数，直接用缩写 ft 即可。

1 ft = 0.3048 m（米）

Exercises

1. Write T (True) or F (False) beside the following statements about the text.

a. With the help of different amplifiers, the users can easily change their location with disruption in their network access.

b. Wi-Fi uses radio networks to transmit data between its nodes.

c. Wi-Fi networks don't have limited range.

d. Wi-Fi in the 5 GHz frequency block has better range than Wi-Fi in the 2.4 GHz frequency block, and less range than the oldest Wi-Fi (and pre-Wi-Fi) 900 MHz block.

2. Fill in the missing words according to the text.

a. Wi-Fi provides its users with the liberty of connecting to the Internet from_____ such as their home, office or a public place without the _____ in the wires.

b. _____ can comprise an area as small as a single room with walls that block radio waves or as large as many square miles—this is achieved by _____.

c. Wi-Fi location where the users can connect to the wireless network is called a _____.

d. The more number of cells, the _____ and _____ is the coverage on the radio network.

3. Translate the following paragraph into Chinese.

Wireless Fidelity, popularly known as Wi-Fi, developed on IEEE 802.11 standards, is a widely used technology advancement in wireless communication. As the name indicates, Wi-Fi provides wireless access to applications and data across a radio network. Wi-Fi sets up numerous ways to build up a connection between the transmitter and the receiver such as DSSS, FHSS, IR and OFDM.

Chinese Translation of Texts（参考译文）

第 15 课 无线相容性认证

无线相容性认证，通常称为 Wi-Fi，它由 IEEE 802.11 标准发展而来，是在无线通信中被广泛应用的先进技术。顾名思义，Wi-Fi 对无线网络中的应用程序和数据提供无线接入。Wi-Fi 在发射端和接收端之间设置了许多方法来建立连接，如扩频、跳频、红外的连接和正交频分复用。

Wi-Fi 让用户可以从任何地方（如家中、办公室或公共场所等）自由连接到互联网而不需要插上网线。访问一个大型网络的信息，Wi-Fi 比传统的调制解调器快。借助不同放大器的帮助，用户可以很容易地改变自己的位置而不中断其网络接入。Wi-Fi 设备相互兼容以给用户有效的信息访问。用户可以连接到无线网络的地方被称为 Wi-Fi 热点。在某些情况下，通过热点访问无线网络是免费的，而在另一些地方可能需要额外的费用。许多标准的 Wi-Fi 设备，如 PCI、miniPCI、USB、Cardbus 及 PC 卡、ExpressCard，使得用户的 Wi-Fi 体验方便而愉快。

Wi-Fi 使用无线网络在节点之间传输数据。这种网络是由覆盖到整个网络范围的蜂窝组成的。蜂窝的数量越多，无线网络的覆盖范围越大、越强。最流行的 Wi-Fi 技术，如 802.11b，工作在 2.40 GHz 到 2.4835 GHz 频段。这为蓝牙策略、手机和其他科学设备的工作提供了一个全面的平台。虽然 802.11a 技术有着 5.725 GHz 到 5.850 GHz 的范围，并可提供高达 54 Mbit/s（兆比特/秒）的速度，但 802.11g 技术甚至更好，它包括 3 个非重叠信道，并允许分组广播控制信道。

可以使用无线网络的设备（如个人计算机、游戏机、智能手机、平板电脑或数字音频播放器）都可以通过无线网络接入点连接到互联网等网络资源。Wi-Fi 网络的范围有限。一个使用 802.11b 或 802.11g 的典型 Wi-Fi 家庭路由器可能有 45 米（150 英尺）的室内范围和 90 米（300 英尺）的室外范围。由于 Wi-Fi 毫无例外的是使用无线电波进行传播的，其范围也会随着频段变化而变化。2.4 GHz 频段的 Wi-Fi 的传播范围比 5 GHz 频段的更大，但小于早期 900 MHz 频段的 Wi-Fi。

热点覆盖区域包括小到一个有阻止无线电波的墙的房间，大到几平方英里——这是通过使用多个重叠的接入点实现的（见图 15-1）。

Unit III Communicated Technology 通信技术

Lesson 16 Global Positioning System (GPS)

GPS is the Global Positioning System. GPS uses satellite technology to enable a terrestrial terminal to determine its position on the Earth in latitude and longitude.

GPS receivers do this by measuring the signals from three or more satellites simultaneously and determining their position using the timing of these signals (Fig.16-1).

GPS operates using trilateration. Trilateration is the process of determining the position of an unknown point by measuring the lengths of the sides of an imaginary triangle between the unknown point and two or more known points.

Fig.16-1 Global Positioning System (GPS)

In the GPS system, the two known points are provided by two GPS satellites. These satellites constantly transmit an identifying signal. The GPS receiver measures the distance to each GPS satellite by measuring the time each signal took to travel between the GPS satellite and the GPS receiver.

The GPS system is divided into three segments.

- The Space Segment
- The Control Segment
- The User Segment

GPS uses twenty-one operational satellites, with an additional three satellites in orbit as redundant backup. GPS uses NAVSTAR satellites manufactured by Rockwell International. Each NAVSTAR satellite is approximately 5 meters wide (with solar panels extended) and weighs approximately 900 kg.

GPS satellites orbit the earth at an altitude of approximately 20,200 km.

Each GPS satellite has an orbital period of 11 hours and 58 minutes. This means that each GPS satellite orbits the Earth twice each day.

Fig.16-2 GPS Receiver

These twenty-four satellites orbit in six orbital planes, or paths. This means that four GPS satellites operate in each orbital plane. Each of these six orbital planes is spaced sixty degrees apart. All of these orbital planes are inclined fifty-five degrees from the Equator.

In order for GPS tracking to work, it is necessary to have both access to the Global Positioning System and have a GPS receiver (Fig.16-2). The GPS receiver is able to receive signals that are transmitted by GPS satellites orbiting overhead. Once these satellite transmissions are received by the GPS receiver, location and other information such as speed and direction can be calculated.

The receiver contains a mathematical model to account for these influences, and the satellites also broadcast some related information which helps the receiver in estimating the correct speed of propagation. Certain delay sources, such as the ionosphere, affect the speed of radio waves based on their frequencies; dual frequency receivers can actually measure the effects on the signals.

In order to measure the time delay between satellite and receiver, the satellite sends a repeating 1,023 bit long pseudo random sequence; the receiver constructs an identical sequence and shifts it until the two sequences match.

Different satellites use different sequences, which lets them all broadcast on the same frequencies while still allowing receivers to distinguish between satellites. This is an application of Code Division Multiple Access, CDMA.

There are two frequencies in use: 1575.42 MHz (referred to as L1), and 1227.60 MHz (L2). The L1 signal carries a publicly usable coarse-acquisition (C/A) code as well as an encrypted P(Y) code. The L2 signal usually carries only the P(Y) code.

New Words

terrestrial [tə'restriəl] adj.	地球的，地上的
latitude ['lætitju:d] n.	纬度
longitude ['lɔndʒitju:d] n.	经度
trilateration [ˌtrailætə'reiʃən] n.	【测】三边测量（术）
redundant [ri'dʌndənt] adj.	多余的
segment ['segmənt] n.	段；部分
orbit ['ɔ:bit] n. & v.	轨道，常轨；绕轨道而行
delay [di'lei] n. & v.	耽搁，延迟
ionosphere [ai'ɔnəsfiə] n.	电离层
pseudo ['sju:dəu] adj.	假的，冒充的
sequence ['si:kwəns] n.	序列，继起的事，顺序
multiple ['mʌltipl] adj.	多样的，多重的
access ['ækses] n.	通路，进入，使用之权
encrypt [in'kript] v.	加密，将……译成密码

Notes

1. latitude and longitude   经纬度

   例如：One used to represent a unit of measurement, such as feet or minutes in latitude and longitude.

译文：用来代表某一测量单位，如经度或纬度上的英尺或分。

例如：Nothing makes the earth seem so spacious as to have friends at a distance; they make latitude and longitude.

译文：没有一件事情比有朋在远方更使这地球显得如此广大的了；他们构成纬度，也构成经度。

例如：GPS is the Global Positioning System. GPS uses satellite technology to enable a terrestrial terminal to determine its position on the Earth in latitude and longitude.

译文：GPS 是全球卫星定位系统。GPS 使用卫星技术，能使地面终端确定它在地球上的经度和纬度。

2. simultaneously adv. 同时地（联立地）

例如：That cannot be simultaneously true; mutually exclusive.

译文：互相矛盾的，不能同时成立的；互斥的

例如：A group of three bits or three pulses, usually in sequence on one wire or simultaneously on three wires.

译文：3 个二进制位或 3 个脉冲组成的一组，通常按时间顺序出现在一根导线上或同时出现在 3 根导线上。

例如：GPS receivers do this by measuring the signals from three or more satellites simultaneously and determining their position using the timing of these signals.

译文：GPS 接收器同时测量来自 3 个或 3 个以上的卫星的信号，并通过这些信号的传播时间来确定位置。

3. NAVSTAR satellites

导航卫星

4. Rockwell International

罗克韦尔国际公司

例如：GPS uses NAVSTAR satellites manufactured by Rockwell International.

译文：全球定位系统采用由罗克韦尔国际公司制造的导航卫星。

5. equator n. （地球或天球的）赤道；（平分球形物体的面的）圆

the celestial equator: 天球赤道

the earth's equator: 地球赤道

the magnetic equator: 地磁赤道

例如：All of these orbital planes are inclined fifty-five degrees from the equator.

译文：所有这些轨道面都偏离赤道倾斜 55°。

Exercises

1. Write T (True) or F (False) beside the following statements about the text.

a. GPS uses computer technology to enable a terrestrial terminal to determine its position on the Earth.

b. GPS receivers do this by measuring the signals from only one satellites simultaneously and determining their position using the timing of these signals.

c. Trilateration is the process of determining the position of an unknown point by measuring the temperature of the sides of an imaginary triangle between the unknown point and two or more known points.

d. Each NAVSTAR satellite is approximately 50 meters wide and weighs approximately 90 kg.

e. In order for GPS tracking to work, it is necessary to have access to the Global Positioning System but have a GPS receiver

f. In the GPS system, the two known points are provided by one GPS satellites.

g. The GPS receiver is able to receive signals that are transmitted by stars orbiting overhead.

h. Even these satellite transmissions are received by the GPS receiver, location and other information such as speed and direction can not be calculated.

2. Match the following terms to appropriate definition or expression.

a. GPS   1. Space, Control and User Segment

b. CDMA   2. Code Division Multiple Access

c. orbital period of 11 hours and 58 minutes   3. Global Positioning System

d. GPS system is divided into   4. orbits the Earth twice each day

3. Fill in the missing words according to the text.

a. In order to measure the time delay between _____ and _____, the satellite sends a repeating 1,023 bit long pseudo random sequence; the receiver constructs an identical sequence and shifts it until the _____ match.

b. These satellites constantly transmit an identifying _____.

c. The GPS receiver measures the _____ to each GPS satellite by measuring the time each signal took to travel between the _____ and the GPS receiver.

d. GPS uses _____ operational satellites, with an additional _____ in orbit as redundant backup.

4. Translate the following paragraph into Chinese.

The receiver contains a mathematical model to account for these influences, and the satellites also broadcast some related information which helps the receiver in estimating the correct speed of propagation. Certain delay sources, such as the ionosphere, affect the speed of radio waves based on their frequencies, dual frequency receivers can actually measure the effects on the signals.

Chinese Translation of Texts（参考译文）

第16课 全球定位系统

GPS是全球卫星定位系统。GPS使用卫星技术，能使地面终端确定它在地球上的经度和纬度的位置。

接收器同时测量来自3个或3个以上的卫星的信号，并通过这些信号的传播时间来确定位置（见图16-1）。

GPS的工作使用三边测量法。三边测量是指在未知点两侧有两个或两个以上的已知点，与未知点呈虚拟的三角形，通过测量未知点到已知点的边长来确定未知点。

在GPS中，两个已知点由两个全球定位系统卫星提供。这些卫星不断传送一个确定的信号。GPS接收机通过测量每个信号在GPS卫星和GPS接收机之间传播的时间来测量其距离。

全球卫星定位系统分为3个部分：

- 空间部分；
- 控制部分；
- 用户部分。

GPS使用21颗运行卫星，轨道中另外还有3颗后备卫星。全球定位系统采用由罗克韦尔国际公司制造的导航卫星。每一个导航卫星约5 m宽（包括延伸太阳能发电板的长度），大约900 kg。

GPS卫星的轨道离地区的距离约为20200 km。

每个GPS卫星绕轨道一周的时间为11小时58分钟。这意味着每一个GPS卫星每天绕地球两次。

这24颗卫星有6个轨道面，或路径，这意味着每个轨道面有4个GPS卫星在工作。6个轨道面的每个之间呈60°隔开。所有这些轨道面都偏离赤道倾斜55°。

GPS为了进行跟踪，既要访问全球定位系统，还要有一个全球定位系统接收机（见图16-2）。全球定位系统接收机能收到上方绕轨道运行的全球定位系统卫星传送的信号。一旦这些卫星传送的信号被全球定位系统接收机收到，位置及其他信息（如速度和方向等）就能计算出来。

接收机包含一个数学模型来计算这些信息的影响，卫星也传播一些相关的信息以帮助接收机正确地估算传播速度。电离层等一些原因会影响不同频率的无线电波的速度，双频接收机能测量它们对信号的影响。

为了测量卫星和接收机之间的时间延迟，卫星重复发送1023 bit长的伪随机序列；接收机建立一个完全相同的序列并移动它直到两序列匹配。

不同卫星使用不同的序列，这样，当不同的卫星发送相同频率的信号时，接收机能够区分它们。这就是码分复用（CDMA）的应用。

有两个频率GPS被使用：1575.42 MHz（称为L1频率）和1227.60 MHz（L2频率），L1信号携带一个可公开使用的粗码（C/A）和一个加密码P（Y）。L2信号通常只携带P（Y）码。

Lesson 17 Wireless Sensor Network

A Wireless Sensor Network (WSN) consists of spatially distributed autonomous sensors to monitor physical or environmental conditions, such as temperature, sound, pressure, etc., and to cooperatively pass their data through the network to a main location. The more modern networks are bi-directional, also enabling control of sensor activity. The development of wireless sensor networks was motivated by military applications such as battlefield surveillance; today such networks are used in many industrial and consumer applications, such as industrial process monitoring and control, machine health monitoring, and so on.

The WSN is built of "nodes"—from a few to several hundreds or even thousands, where each node is connected to one (or sometimes several) sensors (as shown in Fig.17-1). Each such sensor network node has typically several parts: a radio transceiver with an internal antenna or connection to an external antenna, a microcontroller, an electronic circuit for interfacing with the sensors and an energy source. A sensor node might vary in size from that of a shoebox down to the size of a grain of dust. The cost of sensor nodes is similarly variable, ranging from a few to hundreds of dollars, depending on the complexity of the individual sensor nodes. Size and cost constraints on sensor nodes result in corresponding constraints on resources such as energy, memory, computational speed and communications bandwidth. The topology of the WSNs can vary from a simple star network to an advanced multi-hop Wireless Mesh Network (WMN).

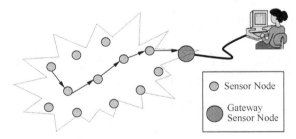

Fig.17-1 Wireless Sensor Network

Applications

Area monitoring

Area monitoring is a common application of WSNs. In area monitoring, the WSN is deployed over a region where some phenomenon is to be monitored. A military example is the use of sensors to detect enemy intrusion; a civilian example is the geo-fencing of gas or oil pipelines.

Environmental/Earth monitoring

The term Environmental Sensor Networks has evolved to cover many applications of WSNs to earth science research. This includes sensing volcanoes, oceans, glaciers, forests, etc. For example, forest fire detection. A network of Sensor Nodes can be installed in a forest to detect when a fire has started. The nodes can be equipped with sensors to measure temperature, humidity and gases which are produced by fire in the trees or vegetation. The early detection is crucial for a successful action of the firefighters; thanks to Wireless Sensor Networks, the fire brigade will be able to know when a fire is started and how it is spreading.

Data logging

Wireless sensor networks are also used for the collection of data for monitoring of environmental information. The statistical information can then be used to show how systems have been working. The advantage of WSNs over conventional loggers is the "live" data feed that is possible.

Industrial sense and control applications

In recent research, a vast number of wireless sensor network communication protocols have been developed. While previous research was primarily focused on power awareness, more recent research have begun to consider a wider range of aspects, such as wireless link reliability, real-time capabilities, or quality-of-service. These new aspects are considered as an enabler for future applications in industrial and related wireless sense and control applications, and partially replacing or enhancing conventional wire-based networks by WSN techniques.

Smart home monitoring

Monitoring the activities performed in a smart home is achieved using wireless sensors embedded within everyday objects forming a WSN. State changes to objects based on human manipulation is captured by the wireless sensors network enabling activity-support services.

New Words

bi-directional [baidi'rekʃ(ə)n(ə)l] adj.	双向的，双向作用的
surveillance [sə'veiləns] n.	监视
transceiver [træn'si:və] n.	收发器
constraint [kən'streint] n.	约束，限制
deploy [di'plɔi] vt.	部署
intrusion [in'tru:ʒn] n.	侵扰
evolve [i'vɔlv] v.	发展，进化
volcano [vɔl'keinəu] n.	火山
glacier ['glæsiə] n.	冰川
humidity [hju:'midəti] n.	湿度
vegetation [vedʒə'teiʃn] n.	植物
fire brigade ['faiə] [bri'geid]	消防队

protocol ['prəutəkɔl] n.　　　　　　　协议

embed [im'bed] vt.　　　　　　　　　把……嵌入

Phrases and Expressions

WSN (Wireless Sensor Network)　　　无线传感器网络

WMN (Wireless Mesh Network)　　　　无线网状网络

Notes

1. A wireless sensor network consists of spatially distributed autonomous sensors to monitor physical or environmental conditions, such as temperature, sound, pressure, etc.

    译文：无线传感器网络（WSN）由空间上分布的自主式传感器组成，用以监控物理或环境条件，如温度、声音、压力等。

    consist of: 由……构成，例如，All electronic computers consist of five units although they are of different kinds.

2. A sensor node might vary in size from that of a shoebox down to the size of a grain of dust.

    译文：传感器节点的尺寸可从一个鞋盒大小到一粒灰尘大小。

    that 代 size。英语中，有时为了避免重复提到过的名词，常用 that 或 those 作替代词。其中，that 替代单数名词，只能指物；those 替代复述名词，既可指物，也可指人。例如，

    （1）The climate of Shenyang is just as bad as that of Beijing.（2）The houses of the rich are generally larger than those of the poor.

3. The term Environmental Sensor Networks has evolved to cover many applications of WSNs to earth science research.

    译文：环境传感器网络这个术语已逐步包含了无线传感器网络在地球科学研究领域中的多种应用。

    （1）cover: 包含，涉及，例如，The talks are expected to cover other topics too.（2）application: 运用，应用，用法为 application (of sth.) (to sth.)，例如，The application of new scientific discoveries to industrial production methods usually increases efficiency.

4. The advantage of WSNs over conventional loggers is the "live" data feed that is possible.

    译文：无线传感器网络的优点是使"活"的数据的传送成为可能。

    advantage: 有利条件，优势，常与介词 over 连用，指"与……相比的长处"，例如，For certain types of work, wood has advantages over plastic.

5. While previous research was primarily focused on power awareness, more recent research have begun to consider a wider range of aspects, such as wireless link reliability, real-time capabilities, or quality-of-service.

    译文：虽然以前的研究主要集中在功能方面，但最近的研究已经开始考虑更广泛的方面，

如无线链路的可靠性、实时性，或服务质量。

while 在本句中表示对比、转折，例如，While some people think his comedy is funny, others find him offensive.

Exercises

1. Write T (True) or F (False) beside the following statements about the text.

a. Sensors are spatially distributed in a WSN to monitor physical or environmental conditions.

b. Nowadays, wireless sensor networks are only used in industrial and consumer applications.

c. The cost of sensor nodes is very expensive.

d. The nodes equipped with sensors in a forest can measure humidity in the trees or vegetation.

e. The detection by the sensors may help the firefighters put out the fire successfully.

f. The WSNs can collect real-time data for monitoring of environmental information.

g. People have developed a large number of wireless sensor network communication protocols recently.

h. The WSN in a smart home can capture any state changes to objects.

2. Fill in the missing words according to the text.

a. The WSN is built of _____ which is _____ to one or several sensors.

b. The _____ of the WSNs can be a star network or an advanced multi-hop _____.

c. The WSN is _____ over a region where some _____ is to be monitored in area monitoring.

d. Wireless sensors, which is used to _____ the activities performed in _____, are _____ within everyday objects.

3. Translate the following paragraph into Chinese.

The WSN is built of "nodes"—from a few to several hundreds or even thousands, where each node is connected to one (or sometimes several) sensors. Each such sensor network node has typically several parts: a radio transceiver with an internal antenna or connection to an external antenna, a microcontroller, an electronic circuit for interfacing with the sensors and an energy source. A sensor node might vary in size from that of a shoebox down to the size of a grain of dust. The cost of sensor nodes is similarly variable, ranging from a few to hundreds of dollars, depending on the complexity of the individual sensor nodes. Size and cost constraints on sensor nodes result in corresponding constraints on resources such as energy, memory, computational speed and communications bandwidth. The topology of the WSNs can vary from a simple star network to an advanced multi-hop wireless mesh network.

Chinese Translation of Texts（参考译文）

第 17 课　无线传感器网络

无线传感器网络（WSN）由空间上分布的自主式传感器组成，用以监控物理或环境条件，如温度、声音、压力等，并通过网络将数据合成到一个主位置。更现代的网络是双向的，也能够控制传感器的活动。无线传感器网络的发展是出于军事上的应用，如战场监视。目前，这样的网络被用于许多工业和消费类应用，如工业过程监测和控制、机器状况监测等。

无线传感器网络的内置"节点"，从几个到几百甚至上千，其中每个节点都连接到一个（有时是多个）传感器。每一个这样的传感器网络节点通常由几部分组成：内置天线或连接外部天线的无线收发信机、微控制器、传感器接口电路和电源。传感器节点的尺寸可从一个鞋盒大小到一粒灰尘大小。传感器节点的成本同样是可变的，从几美元到数百美元不等，这取决于单个传感器节点的复杂性。传感器节点的尺寸和成本限制了其相应的资源，如能源、存储器、运算速度和通信带宽。无线传感器网络的拓扑结构，可以从一个简单的星型网络到复杂的多跳无线网状网络。

应用

区域监测

区域监测是无线传感器网络中一种常见的应用。在小区监控中，无线传感器网络部署在一个区域，对区域中的一些现象进行监测。例如，在军事应用中利用传感器探测到敌人的入侵；在民用中为天然气或石油管道提供地理围栏服务。

环境/地球监测

环境传感器网络这个术语已逐步包含了无线传感器网络在地球科学研究领域中的多种应用。这包括传感火山、海洋、冰川和森林等。例如，它可以用来监测森林火灾。网络传感器节点可以安装在森林中，当火灾开始时就监测到。配备了传感器的节点可以测量由火灾引起的树木或植物的温度、湿度和气体变化。对于消防队员来说，行动能否成功，早期侦测至关重要。通过无线传感器网络，消防队员能够知道火灾是何时发生，以及如何传播的。

数据记录

无线传感器网络也可用于收集环境信息监测的数据，统计信息可以用来显示系统如何工作。相比传统记录仪，无线传感器网络的优点是使"活"的数据的传送成为可能。

工业传感和控制应用

在最近的研究中，大量的无线传感器网络通信协议被开发出来。虽然以前的研究主要集中在功能方面，但最近的研究已经开始考虑更广泛的方面，如无线链路的可靠性、实时性，或服务质量。这些新的方面被认为是未来的工业应用及相关的无线传感和控制应用的促进者，并将部分取代或增强基于有线的传统网络。

智能家居监控

使用嵌入在日常物品内的无线传感器网络对在智能家居中的活动进行监视已经实现。物品因人为操作而发生状态上的变化会被提供活动支持服务的无线传感器网络捕获。

Lesson 18 4G Network

4G is the fourth generation of wireless communications currently being developed for high speed broadband mobile capabilities. It is characterized by higher speed of data transfer and improved quality of sound. Although not yet defined by the ITU (International Telecommunications Union), the industry identifies the following as 4G technologies.

- WiMAX (Worldwide Interoperability for Microwave Access)
- 3GPP LTE (3rd Generation Partnership Project Long Term Evolution)
- UMB (Ultra Mobile Broadband)
- FLASH-OFDM (Fast Low-latency Access with Seamless Handoff Orthogonal Frequency Division Multiplexing)

The 4G technology is being developed to meet QoS (Quality of Service) and rate requirements that involve prioritization of network traffic to ensure good quality of services. These mechanisms are essential to accommodate applications that utilize large bandwidth such as the following (as shown in Fig.18-1).

Wireless Broadband Internet Access, MMS (Multimedia Messaging Service), Video Chat, Mobile Television, HDTV (High Definition TV), DVB (Digital Video Broadcasting), Real Time Audio, High Speed Data Transfer.

The goal set by ITU for data rates of WiMAX and LTE is to achieve 100Mbit/s when the user is moving with high speed relative to the base station, and 1Gbit/s for fixed positions.

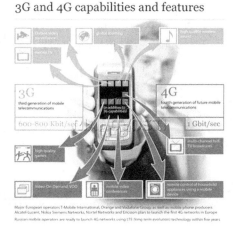

3G and 4G capabilities and features

Fig.18-1 4G network

The industry moves towards expansion of the number of 4G compatible devices. It is set to find its way to tens of different mobile devices not restricted to 4G phones or laptops, such as Video Camera, Gaming Devices, Vending Machines, and Refrigerators.

The trend is to provide wireless internet access to every portable device that could supply and incorporate the 4G embedded modules. The 4G technology could not only provide internet broadband connectivity but also a high level of security that is beneficial to devices that incorporate financial transactions such as vending machines and billing devices.

The following key features can be observed in all suggested 4G technologies.

- MIMO: to attain ultra-high spectral efficiency by means of spatial processing including multi-antenna and multi-user MIMO.
- Frequency-domain-equalization, for example, multi-carrier modulation (OFDM) in the

downlink or single-carrier.

Frequency-Domain-Equalization (SC-FDE) in the uplink: to exploit the frequency selective channel property without complex equalization.

- Turbo principle error-correcting codes: to minimize the required SNR at the reception side.
- Channel-dependent scheduling: to use the time-varying channel.
- Link adaptation: adaptive modulation and error-correcting codes.
- Mobile-IP utilized for mobility.
- IP-based femtocells (home nodes connected to fixed Internet broadband infrastructure).

New Words

interoperability ['intər,ɔpərə'biləti] n.	互通性，互操作性
latency ['leitənsi] n.	潜伏时间，延迟时间
seamless ['si:mləs] adj.	无缝的，不停顿的
handoff ['hændɔf] n.	切换
accommodate [ə'kɔmədeit] vt.	使适应，容纳
relative ['relətiv] adj.	相对的
compatible [kəm'pætəbl] adj.	兼容的
incorporate [in'kɔ:pəreit] v.	包含，使并入
transaction [træn'zækʃn] n.	交易
femtocell ['femtəusel] n.	家庭基站

Phrases and Expressions

4G	第四代移动电话通信标准
ITU	国际电信联盟
WiMAX	全球微波互通存取
3GPP	第三代合作伙伴项目
LTE	长期演进技术
UMB	超移动宽带
OFDM	正交频分复用技术
MMS	多媒体短信服务
HDTV	高清电视
DVB	数字视频广播
MIMO (Multi-Input Multi-Output)	多输入多输出技术
SC-FDE	单载波频域均衡
SNR or S/N (Signal-to-Noise Ratio)	信噪比

spectral efficiency 频谱效率
adaptive modulation 自适应调制

Notes

1. 4G is the fourth generation of wireless communications currently being developed for high speed broadband mobile capabilities.

   译文：4G 是目前正在开发的具有高速移动宽带能力的第四代无线通信技术。

   being developed 为分词结构作定语修饰主语 4G。

2. The 4G technology is being developed to meet QoS (Quality of Service) and rate requirements that involve prioritization of network traffic to ensure good quality of services.

   译文：正在开发的 4G 技术可以满足 QoS（服务质量）和网络流量优先化的速率要求，以保证良好的服务质量。

   meet: 满足，常与 requirement，standard，need 等搭配。例如，(1) How can we best meet the needs of all the different groups? (2) I cannot possibly meet that deadline.

3. The goal set by ITU for data rates of WiMAX and LTE is to achieve 100Mbit/s.

   译文：由 ITU 为 WiMAX 和 LTE 设定的目标是，当用户相对于基站高速移动时，数据速率会达到 100 Mbit/s。

   set by… 为分词作定语修饰 goal。

4. The 4G technology could not only provide internet broadband connectivity but also a high level of security that is beneficial to devices that incorporate financial transactions such as vending machines and billing devices.

   译文：4G 技术不仅可以提供宽带互联网连接，而且具有更高级别的安全性，有利于包含金融交易的设备，如自动售货机和计费装置。

   （1）not only… but (also)…：不仅……而且……，如 They not only talked but also shouted and laughed.（2）that is beneficial…和 that incorporate…为定语从句分别修饰 a high level of security 和 devices。

Exercises

1. Write T (True) or F (False) beside the following statements about the text.

a. 4G is characterized by higher speed of data transfer and improved quality of sound.

b. FLASH-OFDM is regarded as one of 4G technologies.

c. 4G technologies are developed for the applications that require large bandwidth.

d. When the user is in a fixed position, the data rates of WiMAX should achieve 100 Mbit/s in a 4G network.

e. 4G compatible devices only refer to 4G phones or laptops.

f. The 4G technology is beneficial to devices that incorporate financial transactions because it provides a high level of security.

2. Fill in the missing words according to the text.

a. 4G is currently developed for high speed _____.

b. 4G technologies mentioned in the text are _____, _____, _____, _____.

c. ITU set a goal for data rates of WiMAX and LTE is to _____ when the user is _____ relative to the base station, and 1 Gbit/s for _____.

d. The 4G technology could provide _____ and a high level of _____ that is beneficial to devices which _____.

3. Translate the following paragraphs into Chinese.

The 4G technology is being developed to meet QoS (Quality of Service) and rate requirements that involve prioritization of network traffic to ensure good quality of services. These mechanisms are essential to accommodate applications that utilize large bandwidth.

The trend is to provide wireless internet access to every portable device that could supply and incorporate the 4G embedded modules. The 4G technology could not only provide internet broadband connectivity but also a high level of security that is beneficial to devices that incorporate financial transactions such as vending machines and billing devices.

Chinese Translation of Texts（参考译文）

第18课　4G 网络

4G 是目前正在开发的具有高速移动宽带能力的第四代无线通信技术。它的特点是更高的数据传输速率和声音质量。虽然 ITU（国际电信联盟）尚未给出确切的定义，但行业已经确定了以下的 4G 技术。

- WiMAX（全球微波互联接入）
- 3GPP LTE（第三代合作伙伴长期演进项目）
- UMB（超移动宽带）
- FLASH-OFDM（快速低延迟访问的无缝切换正交频分复用）

正在开发的 4G 技术可以满足 QoS（服务质量）和网络流量优先化的速率要求，以保证良好的服务质量。对于使用大带宽的应用来说，这些机制十分重要，如无线宽带上网、MMS（多媒体信息服务）、视频聊天、移动电视、HDTV（高清晰度电视）、DVB（数字视频广播）、实时音频、高速数据传输。

由 ITU 为 WiMAX 和 LTE 设定的目标是，当用户相对于基站高速移动时，数据速率会达到 100 Mbit/s；位置固定时，会达到 1 Gbit/s。

行业的发展带来了大量的 4G 兼容设备。目前已有数十种不同的移动设备，而且不局限于 4G

手机或笔记本电脑，如摄录像机、游戏装置、自动售货机和冰箱等。

目前的趋势是给每一个提供和纳入 4G 嵌入式模块的便携式装置提供无线互联网接入。4G 技术不仅可以提供宽带互联网连接，而且具有更高级别的安全性，有利于包含金融交易的设备，如自动售货机和计费装置。

在所提到的 4G 技术中可以观察到以下几个主要特点。

- 多输入多输出：通过多天线和多用户多入多出的空间处理手段达到超高的频谱效率
- 频域均衡，如在下行链路中的多载波调制（正交频分复用）或上行链路中的单载波频域均衡：利用频率选择性信道特性，而且没有复杂的均衡
- Turbo 原理的纠错码：尽量减少对接收端信噪比的要求
- 基于无线信道的调度：采用时变信道
- 链路适配：自适应调制和纠错码
- 利用移动 IP 的移动性
- 基于 IP 架构的家庭基站（连接到固定网络宽带基础设施的家庭节点）

Reading Material

13. Mobile Communications

Cordless Telephone Systems

Cordless telephone systems are full duplex communication systems that use radio to connect a portable handset to a dedicated base station, which is then connected to a dedicated telephone line with a specific telephone number on the Public Switched Telephone Network (PSTN). In first generation cordless telephone systems (manufactured in the 1980's), the portable unit communicates only to the dedicated base unit and only over distances of a few tens of meters.

Early cordless telephones operate solely as extension telephones to a transceiver connected to a subscriber line on the PSTN and are primarily for in-home use.

Second generation cordless telephones have recently been introduced which allow subscribers to use their handsets at many outdoor locations within urban centers such as London or Hong Kong. Modern cordless telephones are sometimes combined with paging receivers so that a subscriber may first be paged and then respond to the page using the cordless telephone. Cordless telephone systems provide the user with limited range and mobility, as it is usually not possible to maintain a call if the user travels outside the range of the base station. Typical second generation base stations provide coverage ranges up to a few hundred meters. Fig.R13-1 shows cordless telephone examples.

Fig.R13-1　Cordless Telephone

Cellular Telephone Systems

A cellular telephone system provides a wireless connection to the PSTN for any user location within the radio range of the system. Cellular systems accommodate a large number of users over a large geographic area, within a limited frequency spectrum. Cellular radio systems provide high quality service that is often comparable to that of the landline telephone systems. High capacity is achieved by limiting the coverage of each base station transmitter to a small geographic area called a cell so that the same radio channels may be reused by another base station located some distance away. A sophisticated switching technique called a handoff enables a call to proceed uninterrupted when the user moves from one cell to another.

A basic cellular system consists of mobile stations, base stations and a mobile switching center (MSC)(as shown in Fig.R13-2). The Mobile Switching Center is sometimes called a Mobile Telephone

Switching Office (MTSO), since it is responsible for connecting all mobiles to the PSTN in a cellular system. Each mobile communicates via radio with one of the base stations and may be handed-off to any number of base stations throughout the duration of a call. The mobile station contains a transceiver, an antenna, and control circuitry, and may be mounted in a vehicle or used as a portable hand-held unit. The base stations consist of several transmitters and receivers which simultaneously handle full duplex communications and generally have towers which support several transmitting and receiving antennas. The base station serves as a bridge between all mobile users in the cell and connects the simultaneous mobile calls via telephone lines or microwave links to the MSC. The MSC coordinates the activities of all of the base stations and connects the entire cellular system to the PSTN. A typical MSC handles 100,000 cellular subscribers and 5,000 simultaneous conversations at a time, and accommodates all billing and system maintenance functions, as well. In large cities, several MSCs are used by a single carrier.

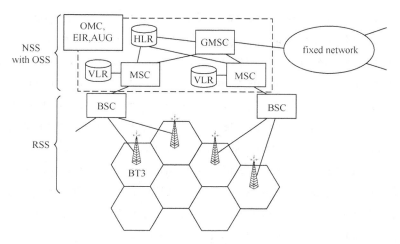

Fig.R13-2 Basic Cellular System

14. Broadband Communications

As can be inferred from the examples of video phone and HDTV, the evolution of future communications will be via broadband communication centered around video signals. The associated services make up a diverse set of high-speed and broadband services ranging from video services such as video phone, video conferencing, video surveillance, Cable Television (CATV) distribution, and HDTV distribution to the high-speed data services such as high-resolution image transmission, high-speed data transmission, and color facsimile. The means of standardizing these various broadband communication services so that they can be provided in an integrated manner is no other than the Broadband Integrated Services Digital Network (B-ISDN). Simply put, therefore, the future communications network can be said to be a broadband telecommunication system based on the B-ISDN (as shown in Fig.R14-1).

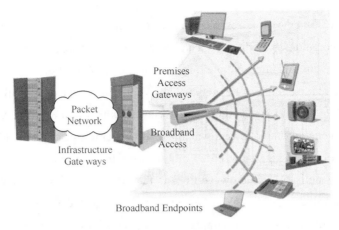

Fig.R14-1　Broadband System and Endpoints

For realization of the B-ISDN, the role of several broadband communication technologies is crucial. Fortunately, the remarkable advances in the field of electronics and fiber optics have led to the maturation of broadband communication technologies. As the B-ISDN becomes possible on the optical communication foundation, the relevant manufacturing technologies for light-source and passive devices and for optical fiber have advanced to considerable levels. Advances in high-speed device and integrated circuit technologies for broadband signal processing are also worthy of close attention. There has also been notable progress in software, signal processing, and video equipment technologies. Hence, from the technological standpoint, the B-ISDN has finally reached a realizable state.

On the other hand, standardization activities associated with broadband communication have been progressing. The Synchronous Optical Network (SONET) standardization centered around the T1 committee eventually bore fruit in the form of the Synchronous Digital Hierarchy (SDH) standards of the International Consultative Committee in Telegraphy and Telephony (CCITT), paving the way for synchronous digital transmission based on optical communication. The standardization activities of the Integrated Services Digital Network (ISDN), which commenced in early 1980s with the objective of integrating narrowband services, expanded in scope with the inclusion of broadband services, leading to the standardization of the B-ISDN in late 1980s and establishing the concept of Asynchronous Transfer Mode (AMT) communication in process. In addition, standardization of various video signals is becoming finalized through the cooperation among such organizations as CCITT, the International Radio communications Consultative Committee (CCIR), and the International Standards Organization (ISO), and reference protocols for high-speed packet communication are being standardized through ISO, CCITT, and the Institute of Electrical and Electronics Engineer (IEEE).

Various factors such as these have made broadband communication realizable. Therefore, the 1990s is the decade in which matured broadband communication technologies will be used in conjunction with broadband standards to realize broadband communication networks. In the broadband communication network, the fiber optic network will represent the physical medium for implementing broadband communication, while synchronous transmission will make possible the transmission of broadband service signals over the optical medium. Also, the B-ISDN will be essential as the broadband telecommunication network established on the basis of optical medium and synchronous transmission and ATM is the communication means that enables the realization of the B-ISDN. The most important of the broadband services to be provided through the B-ISDN are high-speed data communication services and video communication services.

15. Router

A network device that forwards packets from one network to another. Based on internal routing tables, routers read each incoming packet and decide how to forward it. The destination address in the packets determines which line (interface) outgoing packets are directed to. In large-scale enterprise routers, the current traffic load, congestion, line costs and other factors determine which line to forward to.

Most routers in the world sit in homes and small offices and do nothing more than direct Web, e-mail and other Internet transactions from the local network to the cable or DSL modem, which is connected to the ISP and Internet (as shown in Fig.R15-1). Sitting at the edge of the network, they often contain a built-in firewall for security, and the firewall serves all users in the network without requiring that the personal firewall in each computer be turned on and configured. See firewall and personal firewall.

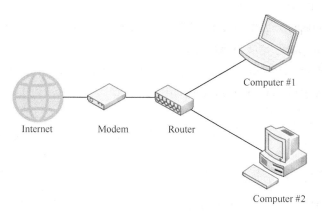

Fig.R15-1   Router

However, in the larger company, routers are also used to separate Local Area Networks (LANs) into subnetworks (subnets) in order to balance traffic within workgroups and to filter traffic for security purposes and policy management.

Within a large enterprise, routers serve as an internet (lower case "i") backbone that connects all internal networks, in which case they are typically connected via Ethernet. Within the global Internet (upper case "I"), routers do all the packet switching between the backbones and are typically connected via T3, ATM or SONET links. See collapsed backbone.

Routers route messages transmitted only by a routable protocol such as IP or IPX. Multiprotocol routers support more than one protocol; for example, IP "and" IPX. Messages in non-routable protocols, such as NetBIOS and LAT, cannot be routed, but they can be transferred from LAN to LAN via a bridge.

Because routers have to inspect the network address in the packet, they do more processing and add more overhead than a bridge or switch. Routers work at the network layer (layer 3) of the protocol, whereas bridges and switches work at the data link layer (layer 2), also known as the "MAC layer." See OSI model.

Most routers are specialized computer-based devices optimized for communications; however, router functions can also be implemented by adding software to a server. For example, NAT32 is software from Microsoft that enables a PC to function as a router to the Internet for machines on the network. The major router vendors are Cisco Systems and Nortel Networks.

Routers used to be called "gateways", which is why the term "default gateway" means the router in your network (see default gateway). In older Novell terminology, routers were also called "network-layer bridges". For more details on the routable protocol layer (network layer 3), see OSI model and TCP/IP abc's. See layer 3 switch, route server, router cluster and routing protocol.

16. Cable Television

Cable television is a system of distributing television programs to subscribers via Radio Frequency (RF) signals transmitted through coaxial cables or light pulses through fiber-optic cables (as shown in Fig.R16-1). This contrasts with traditional broadcast television (terrestrial television) in which the television signal is transmitted over the air by radio waves and received by a television antenna attached to the television. FM radio programming, high-speed Internet, telephone service, and similar non-television services may also be provided through these cables.

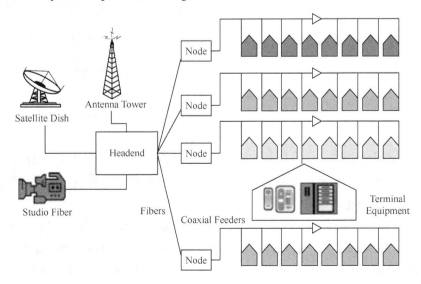

Fig.R16-1　Cable Television System

In order to receive cable television at a given location, cable distribution lines must be available on the local utility poles or underground utility lines. Coaxial cable brings the signal to the customer's building through a service drop, an overhead or underground cable. If the subscriber's building does not have a cable service drop, the cable company will install one. The cable company's portion of the wiring usually ends at a distribution box on the building exterior, and built-in cable wiring in the walls usually distributes the signal to jacks in different rooms to which televisions are connected. Multiple cables to different rooms are split off the incoming cable with a small device called a splitter.

To receive digital cable, most TVs require a digital television adapter (set-top box or cable converter box) from the cable company. A cable from the jack in the wall is attached to the input of the box, and an output cable from the box is attached to the "Antenna In" or "RF In" connector on the back of the television. Different converter boxes are required for newer digital HDTV TVs and older legacy analog televisions. The box must be "activated" by a signal from the cable company before use.

In the most common system, multiple television channels (as many as 500) are distributed to subscriber residences through a coaxial cable, which comes from a trunkline supported on utility poles originating at the cable company's local distribution facility, called the headend. Multiple channels are transmitted through the cable by a technique called frequency division multiplexing. At the headend, each television channel is translated to a different frequency. By giving each channel a different frequency "slot" on the cable the separate television signals do not interfere. At the subscriber's residence, either the subscriber's television or a set-top box provided by the cable company translates the desired channel back to its original frequency (baseband), and it is displayed on the screen.

17. Internet Protocol Television (IPTV)

Internet Protocol Television (IPTV) is a system through which television services are delivered using the Internet protocol suite over a packet-switched network such as the Internet, instead of being delivered through traditional terrestrial, satellite signal, and cable television formats.

IPTV services may be classified into three main groups.
- live television, with or without interactivity related to the current TV show.
- time-shifted television: catch-up TV (replays a TV show that was broadcast hours or days ago), start-over TV (replays the current TV show from its beginning).
- Video On Demand (VOD): browse a catalog of videos, not related to TV programming.

IPTV is distinguished from Internet television by its on-going standardization process (e.g., European Telecommunications Standards Institute) and preferential deployment scenarios in subscriber-based telecommunications networks with high-speed access channels into end-user premises via set-top boxes or other customer-premises equipment.

IPTV is defined as the secure and reliable delivery to subscribers of entertainment video and related services. These services may include, for example, Live TV, Video On Demand (VOD) and Interactive

TV (iTV). These services are delivered across an access agnostic, packet switched network that employs the IP protocol to transport the audio, video and control signals(as shown in Fig.R17-1). In contrast to video over the public Internet, with IPTV deployments, network security and performance are tightly managed to ensure a superior entertainment experience, resulting in a compelling business environment for content providers, advertisers and customers alike.

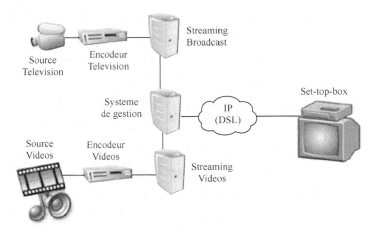

Fig.R17-1   Internet Protocol Television (IPTV)

The Internet protocol-based platform offers significant advantages, including the ability to integrate television with other IP-based services like high speed Internet access and VoIP.

A switched IP network also allows for the delivery of significantly more content and functionality. In a typical TV or satellite network, using broadcast video technology, all the content constantly flows downstream to each customer, and the customer switches the content at the set-top box. The customer can select from as many choices as the telecommunications, cable or satellite company can stuff into the "pipe" flowing into the home. A switched IP network works differently. Content remains in the network, and only the content the customer selects is sent into the customer's home. That frees up bandwidth, and the customer's choice is less restricted by the size of the "pipe" into the home. This also implies that the customer's privacy could be compromised to a greater extent than is possible with traditional TV or satellite networks. It may also provide a means to hack into, or at least disrupt (see Denial of service) the private network.

IPTV technology is bringing Video On Demand (VOD) to television, which permits a customer to browse an online program or film catalog, to watch trailers and to then select a selected recording. The playout of the selected item starts nearly instantaneously on the customer's TV or PC.

Technically, when the customer selects the movie, a point-to-point unicast connection is set up between the customer's decoder (set-top box or PC) and the delivering streaming server. The signalling for the trick play functionality (pause, slow-motion, wind/rewind etc.) is assured by RTSP (Real Time Streaming Protocol).

18. Integration of Three Networks

Three-network integration, namely, the integration of telecommunications, broadcasting and Internet networks (as shown in Fig.R18-1), has been highlighted as national strategy. Industries covered by three-network integration, including the broadcasting industry, the telecommunication industry and the Internet industry, are all tech-intensive and have a sound application basis in China. With a tremendous industry scale, they are important components of the electronic information industry. Therefore, promotion of three-network integration complies with the strategy of economic restructuring.

Fig.R18-1   Integration of Three Networks

Experts expressed that the three-network integration would effectively promote the transformation and upgrade of the existing traditional industries, enable companies to launch various hardware products relating to the three-network integration. Meanwhile, the cultural and creative industries including broadcast, television, cartoon and originality will actively develop and grow with the favorable conditions of the three-network integration and produce and form more new cultural and creative products matching with the pattern of the three-network integration.

The challenges facing the convergence

(1) Monopoly issues. Monopoly issues have occurred in both telecoms and broadcasting industries. Broadcasters have long had a duty to self-regulate their content. On the other hand, telecoms monopolies in broadband networks have hindered other players from gaining access. Currently the number of China's broadband clients is over 1000 million, but telecoms operators occupy over 97% of the market with only 2% from within the broadcasting industry, a huge contrast when compared with the American situation where the ratio for broadcasters and broadband operators is 4:3.

Telecommunication players are positive in predicting the advantages of increased bandwidth and the marketing experience of broadcasters. A survey shows that most people are receptive to the value of services delivered by telecoms operators. Forty percent of mobile subscribers are willing to pay higher fees for better services, such as games, TV, or music on the go. Some popular blog sites, or user-created content-hosting portals, are welcoming 10,000 to 20,000 new subscribers every single day.

(2) Regulatory issues

Given that telecoms, broadcasting and Internet industries are subject to different regulatory bodies, breaking down the monopoly in their own territory is hard work for regulators. Some believe that an umbrella regulator, that would take charge of the country's telecom and media industries, should be constructed along the lines of the US Federal Communications Commission. "A special work party to coordinate and supervise different departments involved during the convergence is needed." Many industry analysts, however, believe convergence and the establishment of a new regulator will not take place until after a new telecoms law is issued.

(3) Technological issues

On the surface convergence only became possible as a result of fundamental changes in technology of the past decade, which have transformed the market dynamics of the telecommunication, media, Internet, hardware, and software industries. The development of new technologies has enabled the fixed and wireless worlds to come together. These technical developments are making possible the transport of high-quality/high-definition audio and video streams onto IP networks accessible from both fixed and wireless devices. Some experts, however, worry more about technical standards in China. A big contrast can be seen between the telecommunication and broadcasting sectors. While the former has benefited from an open attitude towards technical standards and flexible marketing models, the latter suffered from confusion regarding these standards. The different standards have contributed to the failure of the digitalization of TV undertaken by broadcasters, leaving marketing much less dynamic than telecommunication. Even in the telecoms sector where bandwidth is traditionally strong, more investment is still desperately needed to adjust the network appropriately. Thus with both industries planning to reconstruct the network waste is inevitable, and what's more worrying is that on most occasions it is the consumers who will pay for that waste.

Unit IV

Advanced Electronic Technology

电子高新技术

本章将重点介绍应用广泛的高新电子技术，主要包括物联网、计算机仿真、遥感、人工智能、二维码、光伏设备、数字图像处理、计算机视觉、多媒体技术和射频识别等。通过本章的学习，学生可以了解当前发展迅速的高新电子技术，开阔视野，丰富专业知识，并学会阅读具有一定广度和深度的英文技术资料。

Lesson 19 Internet of Things

Internet of Things (IoT), also known as the sensor network, refers to a variety of information sensing devices and the Internet combine to form a huge network, will enable all of the items and network connections to facilitate the identification and management. Because of its comprehensive sense, reliable delivery, intelligent processing features, it is considered as another wave of the information industry after the computer, the Internet and mobile communication network.

Touch of a button on the computer or cell phone, even thousands of miles away, you can learn the status of an item, a person's activities. Send a text message, you can turn on the fan; if an illegal invasion of your home takes place, you will receive automatic telephone alarm. They are not just the scenes in Hollywood sci-fi blockbusters. They are gradually approaching in our lives.

It can be achieved due to the "things" in which there is a key technology for information storage object called Radio Frequency Identification (RFID). An RFID system consists of three components (as shown in Fig.19-1): an antenna, a transceiver (often combined into a reader) and a transponder (the tag). The antenna emits radio signals to activate the tag and to read and write data to it. When activated, the tag transmits data back to the antenna. The data transmitted by the tag may provide identification or location information, or specifics about the product tagged, such as price, color, date of purchase, etc. Low-frequency RFID systems (30 kHz to 500 kHz) have short transmission ranges (generally less than six feet). High-frequency RFID systems (850 MHz to 950 MHz and 2.4 GHz to 2.5 GHz) offer longer transmission ranges (more than 90 feet). In general, the higher the frequency, the more expensive the system.

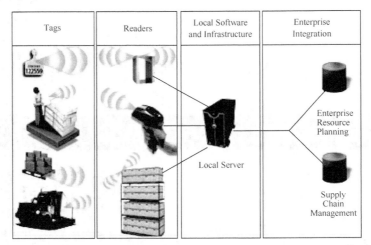

Fig.19-1　A RFID System

For example, in mobile phones, embedded RFID-SIM card, your phone "information sensing device" can be connected with the mobile network. This phone can not only confirm the user's identity, but also to pay the bills for water, gas and electricity, lottery, airline tickets and other payment services.

As long as an object embedded in a specific radio frequency tags, sensors and other devices connected to the Internet will be able to form a large network systems. On this line, even thousands of miles away, people can easily learn and control the information of the object (as shown in Fig.19-2).

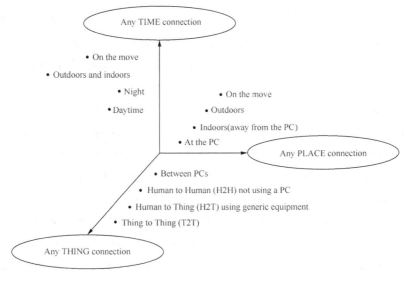

Fig.19-2　How will Internet of Things Affect Our Lives?

Unit Ⅳ　Advanced Electronic Technology 电子高新技术

To speak more concretely, let's imagine a world in which a large number of things that surround us are "autonomous", because they have:

a name, a tag with a unique code;

a memory, to store everything that they cannot obtain immediately from the net;

a means of communication, mobile and energy-efficient, if possible;

sensors, in order to interact with their environment;

acquired or innate behaviors, to act according to a logic, an objective given by its owner.

And, of course, like everything else on Earth, these things must have an electronic existence on the network.

Some experts predict that, in 10 years, "things" may become very popular, and develop into a trillion-scale high-tech market. Then, in almost all areas, such as the personal health, traffic control, environmental protection, public safety, industrial monitoring, elderly care, "things" will play a role. Some experts said that in only three to five years' time, it will change people's way of life.

The Internet of Things has great promise, yet business, policy, and technical challenges must be tackled before these systems are widely embraced. Early adopters will need to prove that the new sensor driven business models create superior value. Industry groups and government regulators should study rules on data privacy and data security, particularly for uses that touch on sensitive consumer information. On the technology side, the cost of sensors and actuators must fall to levels that will spark widespread use. Networking technologies and the standards that support them must evolve to the point where data can flow freely among sensors, computers, and actuators. Software to aggregate and analyze data, as well as graphic display techniques, must improve to the point where huge volumes of data can be absorbed by human.

New Words

sensor['sensə] n.	传感器
facilitate[fə'siliteit] vt.	使容易，促进，帮助
identification[ai,dentifi'keiʃn] n.	认出，识别
illegal[i'li:gl] adj.	不合法的，非法的
storage['stɔ:ridʒ] n.	存储，存储器
alarm[ə'la:m] vt.& n.	警告；报警器
bill [bil] n.	账单，目录
transponder [træns'pɔndə] n.	应答器
approaching [ə'prəutʃiŋ] vt.	接近，靠近
expert ['ekspə:t] n.	专家，能手
energy-efficient	能效高的，高能效的
interact [,intər'ækt] v.	交互

121

tackle ['tækl] vt.	拦截
embrace [im'breis] v.	拥抱，包括
spark [spɑ:k] vt.	发动，触发
aggregate ['ægrigət, 'ægrigeit] vt.	使聚集，使积聚，总计达

Phrases and Expressions

Internet of Things	物联网
sensor network	传感网
mobile communication network	移动通信网络
product tagged	产品标签
automatic telephone alarm	自动电话报警
radio frequency identification	射频识别

Notes

1. Low-frequency RFID systems (30kHz to 500kHz) have short transmission ranges (generally less than six feet). High-frequency RFID systems (850 MHz to 950 MHz and 2.4 GHz to 2.5 GHz) offer longer transmission ranges (more than 90 feet).

   译文：低频 RFID 系统（30 kHz 至 500 kHz）的传输距离短（一般小于 6 英尺）。高频 RFID 系统（850 MHz 至 950 MHz 和 2.4 GHz 至 2.5 GHz）的传输距离更长（超过 90 英尺）。

   RFID：是 Radio Frequency Identification 的缩写，即射频识别。RFID 是一种非接触式的自动识别技术，它通过射频信号自动识别目标对象并获取相关数据，识别工作无须人工干预，可工作于各种恶劣环境。

2. For example, in mobile phones, embedded RFID-SIM card, your phone "information sensing device" can, be connected with the mobile network RFID-SIM card.

   译文：例如，在手机里嵌入 RFID-SIM 卡，手机内的"信息传感设备"就能与移动网络相连。

   SIM card：SIM 卡是 Subscriber Identity Model（客户识别模块）的缩写，也称为智能卡、用户身份识别卡，GSM 数字移动电话机必须装上此卡方能使用。

   RFID-SIM card：RFID-SIM 卡的外形尺寸与传统的标准 SIM 卡完全一致，基于 2.4GHz 频段，是一种可实现近/中/远距离无线通信功能的手机智能卡，通信距离为 10～500 cm，单向支持 100 m（数据广播）。RFID-SIM 卡属于 CPU 卡的一种，卡内具备 COS（微型操作系统），区别于逻辑加密卡，具备极高级别的安全管理措施，同时可通过手机读写卡内个性化数据。

Unit IV Advanced Electronic Technology 电子高新技术

Exercises

1. Write T (True) or F (False) beside the following statements about the text.

a. Low-frequency RFID systems have short transmission ranges.

b. In general, the lower the frequency, the more expensive the system.

c. High-frequency RFID systems offer short transmission ranges.

d. When activated, the antenna transmits data back to the tag.

2. Fill in the missing words according to the text.

a. An RFID system consists of three components: _____, _____ and _____.

b. The data transmitted by the tag may provide _____ or _____, or specifics about the product tagged, such as price, date of purchase, etc.

c. Networking technologies and the standards that support them must evolve to the point where data can _____ sensors, computers, and actuators.

d. _____ on the computer or cell phone, even thousands of miles away, you can learn the status of an item, _____.

3. Translate the following paragraph into Chinese.

Internet of Things (IoT), also known as the sensor network, refers to a variety of information sensing devices and the Internet combine to form a huge network, will enable all of the items and network connections to facilitate the identification and management. Because of its comprehensive sense, reliable delivery, intelligent processing features, it is considered as another wave of the information industry after computer, the Internet and mobile communication network.

Chinese Translation of Texts（参考译文）

第19课 物　联　网

物联网又名传感网，是指将各种信息传感设备与互联网结合起来而形成的一个巨大网络，可使所有的物品与网络连接，方便识别和管理。因其具有全面感知、可靠传递和智能处理的特点，它被众人认为是继计算机、互联网和移动通信网之后的又一次信息产业浪潮。

轻触一下计算机或者手机的按钮，即使千里之外，你也能了解到某件物品的状况、某个人的活动情况。发一个短信，你就能打开风扇；如果有人非法入侵你的住宅，你还会收到自动电话报警。这些已不再是好莱坞科幻大片中才有的情形了，物联网正在步步逼近我们的生活。

实现这一切是因为物联网里有一个存储物体信息的关键技术，即射频识别（RFID）。RFID 系统由 3 部分组成（见图 19-1）：天线、收发器（通常集成在一个读入器里）和应答器（标签）。天线发出无线电信号激活标签和读写数据。标签传输的数据可能提供识别或定位信息，或者是有关产品标签的信息，如价格、颜色或购买日期等。低频 RFID 系统（30 kHz 至 500 kHz）的传输距离短（一般小于 6 英尺）。高频 RFID 系统（850 MHz 至 950 MHz 和 2.4 GHz 至 2.5 GHz）的传输距离更长（超过 90 英尺）。通常，频率越高，系统的价格越昂贵。

例如，在手机里嵌入 RFID-SIM 卡，手机内的"信息传感设备"就能与移动网络相连，这种手机不仅可以确认使用者的身份，还具有缴纳水电燃气费、彩票投注、航空订票等多种支付服务。

只要将特定物体嵌入射频标签、传感器等设备，与互联网相连后，就能形成一个庞大的联网系统，在这个网上，即使远在千里之外，人们也能轻松获知和掌控物体的信息（见图 19-2）。

更具体地讲，让我们想象一个世界，大量的事物"自动"地围绕着我们，因为它们有：

一个名字，标签与一个独特的代码；

存储器，存储那些不能立即从网上获得的任何东西；

一种通信方式，移动的和高能效的，如果可能的话；

传感器，为了与环境交互；

获取的或自然的行为，由自身给出的符合逻辑的客观行动。

当然，像地球上的其他事物一样，这些都必须以电子信息的形式存在于一个网络当中。

有专家预测 10 年内物联网就可能大规模普及，发展成为上万亿规模的高科技市场。届时，在个人健康、交通控制、环境保护、工业监测及老人护理等几乎所有领域，物联网都将发挥作用。有专家表示，只需三到五年的时间，物联网就会改变人们的生活方式。

互联网有很大的前途，但这个系统被广泛接受之前，业务、政策和技术的挑战必须解决。早期应用者需证明，这种新的传感器驱动的商业模式，能够创造卓越的价值。行业组织和政府监管机构应该学习数据隐私和数据安全的规则，特别是在使用那些接触到的敏感消费者信息时。在技术方面，传感器和驱动器的成本必须下降至某个水平才能实现广泛的使用。网络技术和标准，必须支持数据可以在传感器、计算机和驱动器之间自由流动。收集和分析数据的软件，以及图形显示技术，必须提高到这样一种地步，大量的数据才可以被人们接收。

Unit Ⅳ Advanced Electronic Technology 电子高新技术

Lesson 20   Computer Simulation

Computer simulation as a powerful analytic tool widely used in scientific research and engineering design demonstrates unrivalled advantages. With computer simulation, scientists and engineers do not have to build real primary prototypes when they observe an known phenomenon, analyse a complex process, design a machine or a building, etc. Computer simulation is particularly significant when the object under study and examination is costly or even impossible to be built into a real model. For example, to study the cause of engine malfunction that has led to a series of supersonic planes crashed, or to examine the impact on passengers when an airplane crashes, researchers may have to repeat simulating the calamities over and over again before they can find out what they need to reach a conclusion. Obviously, these can only be simulated by running computer simulation programs or the like, rather than replicating the tragedies. Another example is engineering design, in which engineers have to try many schemes and parameters before they can come up with a satisfactory design. Using computer simulation programs, engineers can accomplish that iterative process each time by inputting different schemes and parameters into their computer models, rather than building many different real models.

Virtually, computer simulation is based on mathematical models representing the nature of the object under study or examination. The mathematical model comprises a series of equations that depict the inherent processes of the object in mathematical terms. A computer simulation program includes algorithms that are derived from those equations. The outcome of simulation is usually expressed in rather abstract forms, for instance, 2D diagrams curves tables, figures.

Over the past years, computer graphics techniques have helped computer simulation by creating realistic 3D images to depict the object to be simulated and the environment around it or the effect imposed on it. Sophisticated computer simulation package capable of providing real-time interactive moving images have emerged, although still beyond the reach of most industries. Meanwhile, many CAD systems have incorporated visual modules to enable engineers to interactively "walk" through their 3D pseudo models on screen to review their designs and present them to their clients. This convenience is particularly important to architects.

Here is an example. To analyse the distribution of stress in a fuselage when the plane is flying, the computer simulation package will first set up a mathematical model for this specific theme which comprises equations derived from aerodynamics, elasticity, structural mechanics, then implement a series of computations on a simplified structure of the plane, based on the finite element stress analysis, finally, it gives the outcome which will be clusters of curves spread over the simplified structure of the plane, each indicating the locations in the fuselage that suffer stress of a uniform value. The accuracy of the simulation depends on the accuracy of the mathematical model, that is, how closely the model is built to represent the real plane and its environment in terms of mathematics, geometry and mechanics.

Computer simulation seems to be the only choice for analysing disasters, calamities and accidents. Quite a lot of impressive simulations have been made on some potentially threatening disasters and

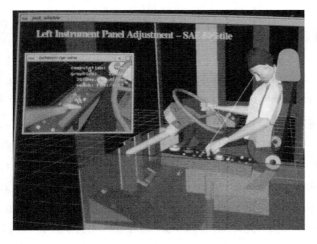

Fig.20-1 Computer Simulation of Bus Operator's Workstation

calamities, for example, the aftermaths of a global nuclear war, the disastrous effect on global climate and environment as a result of worldwide greenhouse effect, the probability of collision between an invading meteorite and the earth, etc.

Other applications of computer simulation technique can be weather forecasting, simulation of combat actions in battlefield, flight programs for training future pilots, human models for testing new medicines and bus operator's workstation (as shown in Fig.20-1).

New Words

simulation [ˌsimju'leiʃn] n.	仿真，假装，模拟
demonstrate ['demənstreit] vt.	示范，证明，论证
prototype ['prəutətaip] n.	原型
phenomenon [fə'nɔminən] n.	现象
malfunction [mæl'fʌŋkʃn] n.	故障
supersonic [ˌsju:pə'sɔnik] adj. & n.	超音波的；超声波
crash [kræʃ] n. & v.	碰撞，坠落，坠毁
impact ['impækt] n.	碰撞，冲击，影响，效果
calamity [kə'læməti] n.	灾难，不幸事件
replicate ['replikit] v.	复制
iterative ['itərətiv] adj.	重复的，反复的
depict [di'pikt] vt.	描述，描写
pseudo ['sju:dəu] adj.	假的，冒充的
architect ['ɑ:kitekt] n.	建筑师
fuselage ['fju:zəlɑ:ʒ] n.	机身
aerodynamics [ˌɛərəudai'næmiks] n.	空气动力学，气体力学
elasticity [ˌelæs'tisəti] n.	弹力，弹性
geometry [dʒi'ɔmətri] n.	几何学
aftermath ['ɑ:ftəmæθ] n.	结果，后果
collision [kə'liʒn] n.	碰撞，冲突
meteorite ['mi:tiərait] n.	陨星
comprise [kəm'praiz] v.	包含，由……组成

Unit Ⅳ Advanced Electronic Technology 电子高新技术

Phrases and Expressions

CAD (Computer Aided Design)　　　计算机辅助设计
greenhouse effect　　　　　　　　　温室效应

Notes

1. Computer simulation as a powerful analytic tool widely used in scientific research and engineering design demonstrates unrivalled advantages.

   译文：计算机仿真作为一种功能强大的分析工具，广泛地用于科学研究和工程设计中，且具有无与伦比的优越性。

   过去分词短语 widely used in scientific research and engineering design 充当后置定语，修饰 computer simulation。

2. For example, to study the cause of engine malfunction that has led to a series of supersonic planes crashed, or to examine the impact on passengers when an airplane crashes, researchers may have to repeat simulating the calamities over and over again before they can find out what they need to reach a conclusion.

   译文：例如，为了研究造成超音速飞机一系列事故的发动机故障的原因，或者为了检查当飞机失事时对乘客的影响，研究者需要一遍又一遍地反复模拟飞机的飞行过程，直到获得最后的结论。

   （1）两个不定式短语充当目的状语，其中，that has led to a series of supersonic planes crashed 是定语从句，修饰 engine malfunction；when an airplane crashes 是时间状语从句。（2）over and over again："一次又一次"，相当于 again and again。（3）before 引导时间状语从句，其中，what 引导宾语从句。（4）find out：v. 计算出，找出，发现，查明（真相等）。

3. Obviously, these can only be simulated by running computer simulation programs or the like, rather than replicating the tragedies.

   译文：显然，他们只能通过计算机模拟程序或其他类似的程序对过程进行仿真模拟，而不是在现实中再制造一场同样的灾难。

   （1）or the like："或者像这种"，指类似于通过操作计算机仿真程序进行模仿这样的事情；
   （2）rather than："而不是"。

4. Sophisticated computer simulation package capable of providing real-time interactive moving images have emerged, although still beyond the reach of most industries.

   译文：虽然还没有广泛应用于工业，但能够提供实时的、交互的运动图像的计算机仿真程序已经出现。

   （1）capable of providing real-time interactive moving images：形容词短语，后置定语，修饰 computer simulation package。（2）在 although still beyond the reach of most industries 中省略

127

了主谓部分 it is，it 指 computer simulation package。

5. …which comprises equations derived from aerodynamics, elasticity, structural mechanics, … finally, it gives the outcome which will be clusters of curves spread over the simplified structure of the plane, each indicating the locations in the fuselage that suffer stress of a uniform value.（1）这里的两个 which 引导定语从句，分别修饰 a mathematical model 和 outcome。（2）it 这里指 computer simulation package。（3）spread over the simplified structure of the plane：过去分词短语，修饰 clusters of curves。（4）each indicating the locations in the fuselage that suffer stress of a uniform value：现在分词的独立结构，充当伴随状语，其中 that 引导定语从句，修饰 locations。

Exercises

1. Write T (True) or F (False) beside the following statements about the text.

a. With computer simulation, scientists and engineers still have to build real primary prototypes.

b. Computer simulation is really useful when making a model costs a lot or even impossible.

c. Even using computer simulation programs, engineers have to try many schemes and parameters before they can come up with a satisfactory design.

d. The technology of computer simulation is based on mathematical models representing the nature of the object under study or examination.

e. The mathematical model can't depict the inherent processes of the object.

f. The outcome of simulation is usually expressed in particular forms.

g. Computer graphics techniques have helped computer simulation by creating realistic 3D images.

h. Computer simulation is the only choice for analysing disasters, calamities and accidents.

2. Fill in the missing words according to the text.

a. Computer simulation as a powerful analytic tool widely used in _____ and _____ demonstrates unrivalled advantages.

b. Other applications of computer simulation technique can be _____, simulation of combat actions in _____, _____ for training future pilots, _____ for testing new medicines.

c. To study the cause of engine malfunction that has led to a series of supersonic planes crashed, or to examine the _____ when an airplane crashes, researchers may have to repeat simulating the tragedies over and over again before they can find out conclusion.

d. Many CAD systems have incorporated _____ to enable engineers to interactively "walk" through their 3D _____ models on screen to review their designs and present them to their clients.

3. Translate the following paragraph into Chinese.

Computer simulation seems to be the only choice for analysing disasters, calamities and accidents. Quite a lot of impressive simulations have been made on some potentially threatening disasters and calamities, for example, the aftermaths of a global nuclear war, the disastrous effect on global climate and environment as a result of worldwide greenhouse effect, the probability of collision between an invading meteorite and the Earth, etc.

Chinese Translation of Texts（参考译文）

第 20 课　计算机仿真

　　计算机仿真作为一种功能强大的分析工具，广泛地用于科学研究和工程设计中，且具有无与伦比的优越性。科学家们在观察已知的现象，分析复杂的过程，设计一台机器或一座建筑物时，利用计算机仿真技术后，不再需要建立真实的原始模型。在研究和考察价格昂贵甚至无法实现的真实模型时，计算机仿真技术显得尤为重要。例如，为了研究造成超音速飞机一系列事故的发动机故障的原因，或者为了检查当飞机失事时对乘客的影响，研究者需要一遍又一遍地反复模拟飞机的飞行过程，直到获得最后的结论。显然，他们只能通过计算机模拟程序或其他类似的程序对过程进行仿真模拟，而不是在现实中再制造一场同样的灾难。另一个例子是在工程设计中工程师必须试验很多方案和参数，才能完成令人满意的设计。利用计算机仿真程序，工程师只需将不同的方案和参数输入到用计算机建立的模型中，就能完成每一次重复过程，而不需建立多个真实模型。

　　实际上，计算机仿真技术是以数学模型为基础的，通过研究和考察后获得的这些数学模型能够代表物体的自然属性。数学模型由一系列描述物体内在过程的，用数学符号表示的数学表达式组成。计算机仿真程序包含了由这些数学表达式获得的算法。仿真的结果通常以相当抽象的形式表达，如二维曲线、图表或数字等。

　　在过去几年里，计算机图形技术已经可以创造真实的三维图像来描述物体，物体周围的环境或其对环境的影响效果，都促进了计算机仿真技术的发展。虽然还没有广泛应用于工业，但能够提供实时的、交互的运动图像的计算机仿真程序已经出现。同时，许多结合了可视模型的计算机辅助设计系统使工程师能够交互地"走"进他们屏幕上的 3D 虚拟模型，检查他们的设计并将它们呈现在客户面前。这种便利对于建筑设计尤为重要。

　　这里还有一个例子，为了分析飞机飞行时机身的应力分布，计算机仿真系统将根据这一特殊主题，建立一个数学模型，数学模型由从空气动力学、弹性力学和结构力学等理论得出的等式组成；然后根据有限元应力分析，针对飞机的简化结构进行一系列计算；最后，由计算机仿真系统算出结果，结果将是一簇遍布飞机的简化结构的曲线，每一条曲线指明了机身上承受同一应力的位置。仿真结果的精确性取决于数学模型的精确性，即数学模型与它代表的真实飞机及用数学、几何学和机械学表示的飞机所在环境的接近程度。

　　计算机仿真似乎是分析灾难、气候和事故的唯一选择。相当多的给人印象深刻的针对具有潜在威胁的灾难的计算机仿真已经被建立，例如，地球核战争的后果、世界范围的温室效应导致的对地球气候和环境的灾难性的影响及天外入侵的陨石与地球相撞的可能性等。

　　计算机仿真还可应用于天气预报、战场上战争情况的模拟、训练飞行员的飞行程序、模仿人类试用新药的反应和汽车驾驶仿真操作等（见图 20-1）。

Lesson 21 Remote Sensing

Remote sensing is the process of collecting data about objects or landscape features without coming into direct physical contact with them. Most remote sensing is performed from aircraft or satellites using instruments, which measure ElectroMagnetic Radiation (EMR) that is reflected or emitted from the terrain(as shown in Fig.21-1). In other words, remote sensing is the detection and measurement of electromagnetic energy emanating from distant objects made of various materials. This is done so that we can identify and categorize these objects by class or type, substance, and spatial distribution.

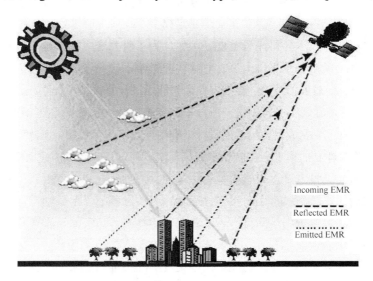

Fig.21-1 Remote Sensing

Remote sensing devices can be differentiated in terms of whether they are active or passive. Active systems, such as radar and sonar, beam artificially produced energy to a target and record the reflected component. Passive systems, including the photographic camera, detect only energy emanating naturally from an object, such as reflected sunlight or thermal infrared emissions. Today, remote sensors, excluding sonar devices, are typically carried on aircraft and earth-orbiting spacecraft.

To complete the remote sensing process, the data captured and recorded by remote sensing systems must be analyzed by interpretive and measurement techniques in order to provide useful information about the subjects of investigation. These techniques are diverse, ranging from traditional methods of visual interpretation to methods using sophisticated computer processing. Accordingly, the two major components of remote sensing are data capture and data analysis.

Thus today we find there are two major branches of remote sensing. The branch first mentioned above is referred to as "image-oriented" because it capitalizes on the pictorial aspects of the data and utilizes analysis methods which rely heavily on the generation of an image. The second branch is referred to as "numerical-oriented" because it results directly from the development of the computer and because it emphasizes the quantitative aspects of the data, treating the data abstractly as a collection of measurement. In this case an image is not thought of as data but rather as a convenient mechanism for viewing the data.

Today we are acquiring earth observational data from earth-orbiting satellites, because of the wide view possible from satellite altitudes, the speed with which the satellite borne sensors travel, and the number of spectral bands used, very large quantities of data are being produced.

Satellite remote sensing may be done two ways.

Using passive sensor systems—Contains an array of small detectors or sensors that can detect electro-magnetic radiations emitted from the Earth's surface.

Using active sensor systems—The system sends out electromagnetic radiation towards target object(s) and measures the intensity of the return signal.

Data collected by the satellites are then transmitted to ground stations wherein images of Earth's surface are reconstituted to obtain the required information.

Take a look at few benefits of satellite remote sensing.

- Enables continuous acquisition of data
- Helps to receive up-to-date information (satellite remote sensing can be programmed to enable regular revisit to object or area under study)
- Offers wide regional coverage and good spectral resolution
- Offers accurate data for information and analysis

New Words

landscape ['lændskeip] n.	风景，山水画，地形，美化
terrain [tə'rein] n.	地面，地域，地带，【军】地形，地势
substance ['sʌbstəns] n.	物质
sophisticated [sə'fistikeitid] adj.	复杂的
radar ['reidɑ:] n.	雷达
sonar ['səunɑ:] n.	声呐
target ['tɑ:git] n.	对象，目标
reflect [ri'flekt] v.	反射（光、热、声或影像）
emanate ['eməneit] vi.	发出，散发，放射
diverse [dai'və:s] adj.	多种多样的
thermal ['θə:ml] adj.	热的，保热的，温热的
array [ə'rei] n.	队列，阵列
intensity [in'tensəti] n.	强烈，强度
reconstitute [,ri:'kɔnstitju:t] vt.	再组成，再构成
acquisition [,ækwi'ziʃn] n.	获得，取得，获得物，【无线】探测

capitalize ['kæpitəlaiz] vi.　　　　　　利用
mechanism ['mekənizəm] n.　　　　　方法，途径，程序，机制，结构

Phrases and Expressions

range from…to…	（范围）从……到……
refer to as	被称为
emanate from	发出，放射
spatial distribution	空间分布
earth-orbiting satellites	地球轨道卫星
image-oriented	图像遥感
result from	产生于……，由……引起
an array of	一排，一群，一批
send out	发送，发出

Notes

1. Most remote sensing is performed from aircraft or satellites using instruments, which measure ElectroMagnetic Radiation (EMR) that is reflected or emitted from the terrain.

   译文：大多数遥感由飞机或卫星携带的仪器完成，这些仪器对从地面发射或反射的电磁辐射（EMR）进行测量。

   EMR：electromagnetic radiation，电磁辐射。

2. Active systems, such as radar and sonar, beam artificially produced energy to a target and record the reflected component. Passive systems, including the photographic camera, detect only energy emanating naturally from an object, such as reflected sunlight or thermal infrared emissions.

   译文：有源系统，如雷达和声呐，对目标发射人为产生的能量，然后记录反射的成分。无源系统，包括照相机，只检测由对象自然反射的能量，如反射的太阳光或热红外辐射。

   active systems：有源系统、主动系统。

   passive systems：无源系统、被动系统。

3. These techniques are diverse, ranging from traditional methods of visual interpretation to methods using sophisticated computer processing.

   译文：这些技术是多种多样的，从传统的视觉判断的方法到复杂的计算机处理。因此，遥感的两个主要组成部分就是数据采集和数据分析。

4. The branch first mentioned above we refer to as "image-oriented" because it capitalizes on the pictorial aspects of the data and utilizes analysis methods which rely heavily on the generation of an image.

   译文：我们把前面首先提到的分支技术称为"图像遥感"，因为它利用图像方面的数据，采用很大程度上依赖于图像生成的分析方法。

Unit Ⅳ Advanced Electronic Technology 电子高新技术

Exercises

1. **Write T (True) or F (False) beside the following statements about the text.**
a. Passive systems beam artificially produced energy to a target and record the reflected component.
b. Active systems detect only energy emanating naturally from an object, such as reflected sunlight or thermal infrared emissions.
c. "Image-oriented" capitalizes on the pictorial aspects of the data and utilizes analysis methods which rely heavily on the generation of an image.
d. "Numerical-oriented" results directly from the development of the computer and because it emphasizes the quantitative aspects of the data, treating the data abstractly as a collection of measurement.
e. Using passive sensor systems—the system sends out electro-magnetic radiation towards target object (s) and measures the intensity of the return signal.

2. **Fill in the missing words according to the text.**
a. Remote sensing is the process of collecting data about _____ without coming into direct physical contact with them.
b. The branch first mentioned above we _____ "image-oriented" because it capitalizes on the _____ aspects of the data and utilizes analysis methods which _____ heavily _____ the generation of an image.
c. Today we are acquiring earth observational data from earth-orbiting satellites, because of the _____ possible from satellite altitudes, the _____ with which the satellite borne sensors travel, and the _____, very large quantities of data are being produced.

3. **Translate the following paragraph into Chinese.**
Remote sensing is the process of collecting data about objects or landscape features without coming into direct physical contact with them. Most remote sensing is performed from aircraft or satellites using instruments, which measure ElectroMagnetic Radiation (EMR) that is reflected or emitted from the terrain. In other words, Remote Sensing is the detection and measurement of electromagnetic energy emanating from distant objects made of various materials. This is done so that we can identify and categorize these objects by class or type, substance, and spatial distribution.

Chinese Translation of Texts（参考译文）

第 21 课　遥感技术

遥感是不与物体直接接触而收集物体或地形特征的有关数据的过程。大多数遥感由飞机或卫星携带的仪器完成，这些仪器对从地面发射或反射的电磁辐射（EMR）进行测量（见图 21-1）。

133

换句话说，遥感可检测和测量远处各种材料的物体所放射的电磁能量。这样做，是为了能够通过类型、物质和空间分布对物体进行识别和分类。

遥感装置可分为有源和无源的。有源系统，如雷达和声呐，对目标发射人为产生的能量，然后记录反射的成分。无源系统，包括照相机，只检测由对象自然反射的能量，如反射的太阳光或热红外辐射。今天，遥感装置，包括声呐设备，通常由飞机和地球轨道航天器携带。

为了完成遥感过程，必须采用编译和测量技术对由遥感系统捕获和记录的数据进行分析以获得关于研究对象的有用信息。这些技术是多种多样的，从传统的视觉判断的方法到复杂的计算机处理。因此，遥感的两个主要组成部分就是数据采集和数据分析。

现今遥感技术有两个主要的分支。我们把前面首先提到的分支技术称为"图像遥感"，因为它利用图像方面的数据，采用很大程度上依赖于图像生成的分析方法。第二个分支称为数字遥感，因为它得益于计算机的发展并强调数据的数量，将测量数据抽象成一个集合。在此情况下，我们并不把图像看作具体的数据来对待，而是把它作为一种便于观察数据的途径。

今天，我们正通过地球轨道卫星上获得的地面观察资料，因为从卫星的高度上能获得宽广的视野，卫星推动传感器运行的速度快，所使用的谱带数目多，产生的数据数量非常大。

卫星遥感可以通过以下两种方式实现。

使用无源传感系统：包括一排小型探测器或传感器阵列，可以检测从地球表面发射的电磁辐射。

使用有源传感系统：系统向一个或多个目标对象发出电磁辐射并测量回波信号的强度。

由卫星收集的数据传输到地面站，在那里，地球表面的图像将被重组以获取所需的信息。

卫星遥感的几个优点如下：

- 能够连续地进行数据采集；
- 有助于获得最新信息（卫星遥感技术可以通过编程来定期对所研究的对象或区域进行重新访问）；
- 提供广泛的区域覆盖和良好的光谱分辨率；
- 提供精确的数据信息和分析。

Lesson 22 Artificial Intelligence (AI)

Since World War II, computer scientists have tried to develop techniques that would allow computers to act more like humans. The entire research effort, including decision-making systems, robotic devices (Fig.22-1), and various approaches to computer speech, is usually called Artificial Intelligence (AI).

An ultimate goal of AI research is to develop a computer system that can learn concepts (ideas) as well as facts, make commonsense decisions, and do some planning. In other words, the goal is to eventually create a "thinking, learning" computer.

A computer program is a set of instructions that enables a computer to process information and solve problems. Most programs are fairly rigid—they tell the computer exactly what to do, step by step. AI programs are, however, exceptions to this rule. They can take short-cuts, make choices, search for and try out different solutions, and change their methods of operation.

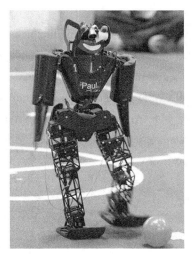

Fig.22-1 Robot

In many AI programs, facts are arranged to enable the computer to tell how many pieces of information relate to each other and to a given problem. "If/then" rules of reasoning are also programmed in to enable the computer to select, organize, and update its information. According to these rules, if something is true, then certain things must follow. Every action makes new possible actions available.

Programs to play chess have been around since the early days of electronic computers, but they tended to be rigid and limited by the skills of the program designer. Detailed instructions on what moves to make and how to respond to an opponent's moves were written into a program. Sometimes the suggestions of several chess experts were included. However, such programs seldom defeated human chess experts. The computer program would tend to be strong in the opening part of the game, but would weaken as the game went on.

Thanks to AI research, all that has changed. Recently, chess-playing computer programs have been developed to defeat most human opponents—including chess masters.

Of course, there's more to artificial intelligence than the ability to play games. Computer scientists are working on dozens of different practical uses for AI programs. These included operating robots, solving math and science problems, understanding speech, and analyzing images.

Perhaps the biggest use of AI programs is expert advisors for trouble-shooting (locating problems and making repairs) complex systems ranging from diesel engines to nuclear submarines and to the human body. In other words, these AI programs search for trouble, detect and classify problem areas, and give advice.

The use of AI expert advice systems will not be limited to trouble-shooting specific machinery. AI programs are being developed for economic planning, weather forecasting, casting, oil exploration, computer design, and numerous other uses.

AI techniques are also being used to analyze human speech and to synthesize speech. With the help of laser sensors, AI techniques are being developed to analyze visual information and to improve robot capabilities.

Most Artificial Intelligence systems involve some sort of integrated technologies, for example the integration of speech synthesis technologies with that of speech recognition. The core idea of AI systems integration is making individual software components, such as speech synthesizers, interoperable with other components, such as common sense knowledge bases, in order to create larger, broader and more capable AI systems. The main methods that have been proposed for integration are message routing, or communication protocols that the software components use to communicate with each other, often through a middleware blackboard system(as shown in Fig.22-2).

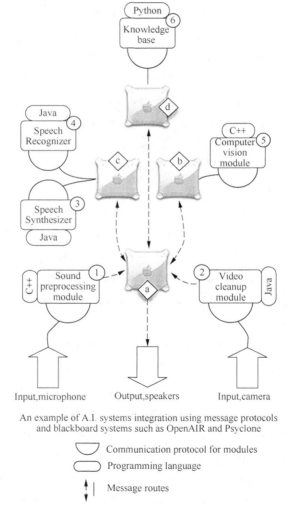

An example of A.I. systems integration using message protocols
and blackboard systems such as OpenAIR and Psyclone

Fig.22-2 Artificial Intelligence Integrated Systems

Unit Ⅳ Advanced Electronic Technology 电子高新技术

New Words

robotic [rəu'bɔtik] n.	机器人的
ultimate ['ʌltimət] adj. & n.	最后的，最终的，根本的；最终
commonsense ['kɔmən'sens] adj.	具有常识的
rigid ['ridʒid] adj.	刚硬的，刚性的，严格的
short-cut ['ʃɔ:tkʌt] n.	捷径
reason ['ri:zn] n. & v.	推理，评理，论证
opponent [ə'pəunənt] n. & adj.	对手，反对者；对立的，对抗的

Phrases and Expressions

trouble-shooting	故障寻找
diesel engine	柴油机
nuclear submarine	核潜艇

Notes

1. The entire research effort, including decision-making systems, robotic devices, and various approaches to computer speech, is usually called Artificial Intelligence (AI).
   译文：所有这些研究，包括决策系统、机器人设备，以及各种各样的计算机语音的实现方法，通常被称为人工智能（AI）。
   现在分词短语 including … to computer speech 充当定语，其作用相当于一个由 which 引导的非限制性定语从句。

2. An ultimate goal of AI research is to develop a computer system that can learn concepts (ideas) as well as facts, make commonsense decisions, and do some planning.
   译文：人工智能的最终目标是开发一种能够从概念及事实中进行学习，做出符合常识的决策，并能做出计划的计算机系统。
   （1）不定式充当表语。（2）that 引导的定语从句修饰 a computer system，从句中有 3 个并列的谓语动词（learn、make 和 do）。

3. … but they tended to be rigid and limited by the skills of the program designer.
   tend to do sth.："倾向于（做）某事，往往做某事"，例如，Children tend to be curious about everything. 再如下文中的 The computer program would tend to be strong in the opening part of the game, but would weaken as the game went on.

4. Computer scientists are working on dozens of different practical uses for AI programs.
   译文：计算机科学家正在为人工智能寻找数十种实际用途。
   dozens：一打，十二个，这里 dozens of 延伸为"数十"，相当于 tens of。

5. With the help of laser sensors, AI techniques are being developed to analyze visual information and to improve robot capabilities.

译文：目前正在开发的人工智能技术还借助于激光传感器分析图像信息，从而提高机器人性能。

（1）with the help of：在……的帮助下，例如，With the help of computers，we can easily carry out some complicated mathematics calculation.（2）两个不定式短语充当目的状语。

Exercises

1. Write T (True) or F (False) beside the following statements about the text.

a. Since World War II, computer scientists have developed techniques that let computers to act same as humans.

b. The goal of AI research is to develop a computer system that can learn concepts, make decisions, do planning and complete all of things for human being.

c. A computer is a "thinking, learning" machine.

d. AI programs are fairly rigid—they tell the computer exactly what to do, step by step.

e. According to the rules, if something is true, then certain nothing will follows.

f. In AI programs, not every action makes new possible actions available.

g. Programs to play chess have no limit and no relate with the skills of the program designer.

h. Computer scientists are working on a lot of different practical uses for AI programs.

2. Match the following terms to appropriate definition or expression.

a. Artificial Intelligence　　　1. to defeat most human opponents—including chess masters

b. a computer program　　　2. to analyze human speech, to synthesize speech, to analyze visual information and to improve robot capabilities

c. AI techniques　　　3. a set of instructions to process information and solve problems

d. chess-playing programs　　　4. decision-making, robotic, and approaches to computer speech

3. Fill in the missing words according to the text.

a. The entire research effort, including _____ systems, _____ devices, and various _____ to computer speech, is usually called Artificial Intelligence (AI).

b. AI programs can take _____, make _____, search for and try out different _____, and change their _____.

c. "If/then" rules of reasoning are also programmed in to enable the computer to _____, _____, and _____ its information.

d. Computer practical uses for AI programs include _____ robots, _____ math and science problems, _____ speech, and _____ images.

4. Translate the following paragraph into Chinese.

Perhaps the biggest use of AI programs is expert advisors for trouble-shooting (locating problems and making repairs) complex systems ranging from diesel engines to nuclear submarines and to the human body. In other words, these AI programs search for trouble, detect and classify problem areas, and give advice.

Unit Ⅳ   Advanced Electronic Technology 电子高新技术

Chinese Translation of Texts（参考译文）

第22课　人　工　智　能

自从第二次世界大战以来，计算机科学家们一直试图开发使计算机能更多地像人一样行动的技术。所有这些研究，包括决策系统、机器人设备（见图 22-1），以及各种各样的计算机语音的实现方法，通常被称为人工智能（AI）。

人工智能的最终目标是开发一种能够从概念及事实中进行学习，做出符合常识的决策，并能做出计划的计算机系统。换句话说，目标是最终造出一台能够"思维和学习"的计算机。

计算机程序是一组使计算机能够处理信息并进行决策的指令。大多数程序相当刻板——它们精确地告诉计算机一步一步做什么。而人工智能程序则不遵守这一规律。它们能够寻找捷径，做出选择，探索和实验不同的答案，并改变解决方案等。

许多人工智能程序把有关事实组织起来，使计算机有可能把不同的信息相互关联，并和某个给定的课题也关联起来。"如果/那么"这一推理规律也被编到程序里去，使计算机能够对信息进行取舍、组织和更新。根据这一规则，若某事为真，另一些事必随之发生，每一步行动必引出一些新的可能的行动。

在电子计算机问世的早期，弈棋程序就已经存在了，但这些程序比较僵化，且受到程序设计人员能力的限制。走哪一步棋、怎样应付对方的棋步等详细的指令都定在程序里。有时程序中也采纳一些象棋专家的建议，但这样的程序很少能战胜象棋高手。计算机程序通常在开局时比较强，但越下越弱。

由于人工智能的出现，这一切都已改变。现在开发的计算机弈棋程序可以击败大多数象棋高手——包括象棋大师。

当然，人工智能不仅仅只会下棋。计算机科学家正在为人工智能寻找数十种实际用途。这包括操纵机器人、解答数学题和科学难题、理解语言、分析图像等。

人工智能程序最大的用途也许是对复杂系统的故障寻迹提供专家咨询（即确定故障的所在并予以修复），所谓复杂系统包括从内燃机到核潜艇到人体。换句话说，这些人工智能程序能够寻找故障、确定故障范围、对故障进行分类，并提供咨询。

人工智能专家系统的应用不限于对特定机械进行故障寻迹。目前正在开发的人工智能程序还可以应用于经济规划、天气预报、石油勘探、计算机设计及许多其他用途。

人工智能技术也用来对人类语言进行分析和合成。目前正在开发的人工智能技术还借助于激光传感器分析图像信息，从而提高机器人性能。

大多数人工智能系统涉及某种形式的集成技术，如语音合成技术和语音识别的集成。人工智能系统集成的核心思想是使单个的软件组件，如语音合成器，可以与其他组件，诸如常识知识库交互操作，以创造更大、更广泛和更强的人工智能系统。集成的主要的方法已经被提出，包括消息路由、软件之间相互沟通的通信协议等，它们往往通过中间件黑板系统进行（见图 22-2）。

Lesson 23   2D Bar Code

Barcodes are a combination of dark and light lines of variable widths. The dark lines are most commonly termed as bars and the light bands as spaces. The elements (bars and spaces) of these barcodes must be consistent, proportional thickness and thinness. What this means is that if the widest elements of the barcode can be as thick as a pencil or as thin as a credit card, the corresponding thin bars and spaces in the barcode remain proportionally thin.

Barcodes are the easiest and the cheapest way of automatic data collection. Instead of an operator reading and punching the product codes by a keyboard at a retail counter, reading a barcode reads the data as well as enters the data into the host system.

A barcode cannot be read by a human eye. Therefore, it is advisable to print the data encoded in a barcode close to the barcode. So the operator can read the data and enter it manually just in case the scanner stops functioning.

Two-dimensional scanners were far more expensive than 1D scanners when were introduced. Recent microprocessor developments have brought the cost of 2D scanners down to about 125% of the cost of a comparable 1D scanner. Also, advancing decoding algorithms have made scanning quicker and easier and provided even greater readability of excessively-damaged symbols. There are a number of 2D symbologies in growing use today. They fall into two categories: matrix and stacked.

2D barcode symbologies represent one of the biggest advances in the market of Automated Data Collection in the past few years. With advancements in technology, with smaller and faster processors, it can only get better. However, when analyzing any potential data collection system, the advantages must be weighed over the added costs. 2D barcode technology should be thought of as one that is complementary to the traditional 1D scanning technology, not its replacement.

Some example applications for 2D technology.

Packing List—Trading partners agree on a standard methodology for encoding shipping information in a 2D symbol, attached to a shipped order. Order data (PO number, shipping date, product codes, quantities, etc.) can automatically be entered into the receiver's receiving computer terminal in a couple of seconds.

Driver's Licence—The driver's name, address, licence number, expiry date and driving restriction codes are encoded in a 2D symbol that is printed on the operator's licence. Police officers, car rental agencies, hotels can easily enter in information regarding the licence holder, with the possibility of adding any mis-keyed data.

Patient Record—On a hospital patient's chart record is a 2D symbol, encoding their name, health care number, doctor's name, date of admission, allergies, etc.

When direct care is given to the patient, the caregiver or doctor records the action by scanning the barcode(as shown in Fig.23-1). Also, the barcode is scanned when medication is administered and the possibility of giving a patient the wrong medicine is virtually eliminated.

Unit IV Advanced Electronic Technology 电子高新技术

Fig.23-1   2D Barcodes

New Words

barcode ['bɑ:kəud] n.	条形码，条码技术
automatic [ɔ:tə'mætik] adj. & n.	自动的，无意识的，必然的；自动机械，自动手枪
symbology [sim'bɔlədʒi] n.	码制；符号学，符号使用，象征学
two-dimensional [ˌtu:di'menʃənəl] adj.	二维的
category ['kætəg(ə)ri] n.	类别，分类
scan [skæn] n. & v.	扫描，浏览，审视

Phrases and Expressions

PO (Purchase Order) number	订单号
computer terminal	计算机终端

Notes

1. The dark lines are most commonly termed as bars and the light bands as spaces.
   译文：黑线条通常称为条，白线条通常称为空。

2. The elements (bars and spaces) of these barcodes must be consistent, proportional thickness and thinness.
   译文：构成条形码的元素（条和空）必须连贯且宽度成比例。

3. So the operator can read the data and enter it manually just in case the scanner stops functioning.
   译文：这样操作者可以读取数据或手动输入，以防扫描仪停止运作。
   in case: 以防，以免，例如，Write the telephone number down in case you forget.

4. 2D scanners were far more expensive than 1D scanners when were introduced. Recent microprocessor developments have brought the cost of 2D scanners down to about 125% of the cost of a comparable 1D scanner.

译文：购买一台二维扫描仪远比一维扫描仪昂贵得多。目前，微处理器的发展已使得二维扫描仪的成本下降至一个同类一维扫描仪成本的125%。

far：非常，（作比较时用于强调）……得多，例如，(1)This English poem is far too hard to translate. (2)There are far more opportunities for young people than there used to be.

5. 2D barcode technology should be thought of as one that is complementary to the traditional 1D scanning technology, not its replacement.

译文：二维码技术应该被看作是传统的一维扫描技术的补充，而不是替代品。

Exercises

1. Write T (True) or F (False) beside the following statements about the text.

a. The elements (bars and spaces) of these barcodes must be consistent, proportional thickness and thinness.

b. A human can read a barcode with their eyes.

c. With a barcode, the operator won't have to enter the data manually.

d. 2D scanners were much cheaper than 1D scanners when were introduced.

e. 1D scanning technology should be replaced by 2D barcode technology.

f. A 2D symbol that is printed on the driver's licence contains the information about the licence holders, such as the name and address.

g. When direct care is given to the patient, the caregiver or doctor records the action by scanning the barcode.

2. Fill in the missing words according to the text.

a. Barcodes are the _____ and the _____ way of automatic data collection.

b. 2D symbologies fall into two categories: _____ and _____.

c. However, when _____ any potential data collection system, the advantages of using 2D barcodes must be _____ the added costs.

d. Order data (_____, shipping date, product codes, quantities, etc.) can _____ be entered into the receiver's receiving _____ in a couple of seconds.

e. On a hospital patient's chart record is a _____, _____ their name, health care number, doctor's name, date of admission, allergies, etc.

3. Translate the following paragraph into Chinese.

A barcode cannot be read by a human eye. Therefore it is advisable to print the data encoded in a barcode close to the barcode generally known as human readable. So the operator can read the data and enter it manually just in case the scanner stops functioning.

Unit Ⅳ Advanced Electronic Technology 电子高新技术

Chinese Translation of Texts（参考译文）

第23课 二维码

条形码是将宽度不同的黑白线条结合起来的一种码。黑线条通常称为条，白线条通常称为空。构成条形码的元素（条和空）必须连贯且宽度成比例。这意味着，如果条形码最宽的条纹像铅笔一样厚或信用卡一样薄，那么最窄的条纹必须与之成比例。

条形码是自动数据采集的最简单和最便宜的方式。操作者无须在零售柜台用键盘输入产品代码，通过读取条形码就能读取产品数据，同时也将数据输入到主机系统。

人眼无法读取条形码，因此，明智的做法是将条形码中被编码的数据打印在靠近条形码的地方，这样操作者可以读取数据或手动输入，以防扫描仪停止运作。

二维码扫描仪诞生之初，购买一台二维扫描仪远比一台一维扫描仪昂贵得多。目前，微处理器的发展已使得二维扫描仪的成本下降至一个同类一维扫描仪成本的125%。此外，推进解码算法使得扫描更快、更容易，甚至为过度损坏的符号提供了更大的可读性。如今，越来越多的二维码制在被使用。它们分为两类：矩阵和堆叠。

二维码的码制是自动数据采集市场在过去的几年中最大的进步之一。随着科技的进步，以及处理器变得更小、更快，它只会变得更好。但是，在分析任何潜在的数据收集系统时，它的优点必须超越其附加成本。二维码技术应该被看作是传统的一维扫描技术的补充，而不是替代品。

2D技术的示例应用如下。

包装清单——贸易伙伴同意用标准方式通过2D符号将装货信息进行编码并置于装船清单上，这样，订单数据（订单号、出货日期、产品代码、数量等）可以在几秒钟内自动进入计算机终端的接收机。

驾驶执照——驾驶员的名称、地址、许可证号、到期日期和驾驶限制编码通过2D符号进行编码，并被印刷在驾驶执照上。在有可能错误键入数据的情况下，警员、租车公司、酒店可以很容易地输入许可证持有人的信息。

病历——医院病人的图表记录上是一个二维的符号，输入的是他们的名字、住院号，医生的姓名，入院日期，过敏史等信息。

当护理病人时，护理人员或医生通过扫描条形码来记录操作（见图23-1）。此外，如果发药时扫描条码，那么给病人发错药的可能性几乎可以完全消除。

Lesson 24 Photovoltaic Technology

The word photovoltaic is a marriage of the words "photo", which means light, and "voltaic", which refers to the production of electricity. Photovoltaics (PV) is the field of technology and research related to the application of solar cells for energy by converting sunlight directly into electricity.

Photovoltaics, or PV, is a green energy source that uses solar radiation in order to generate electric power. It is the most well-known method for turning solar radiation into usable electricity. In photovoltaics, energy from the sun is converted into electrical current using semiconductors. These semiconductors have a photovoltaic effect. During this process, the solar radiation excites electrons, which are then transferred between various bands, such as the conduction bands. This results in a build-up of voltage between the two electrodes in the cell. Photovoltaic technology requires a light source, and primarily depends on energy from the sun.

Photovoltaics are typically seen in the form of solar panels. These solar panels contain a number of solar cells containing photovoltaic material. These panels can be found on roofs of buildings, including homes or businesses. In addition, solar panels can be ground mounted, such as in solar power plants and in the form of photovoltaic trees.

Photovoltaic technology was first used in order to power satellites and spacecraft orbiting in space. However, due to an increase in the availability of photovoltaic technology as well as an increase in the demand for clean energy sources, solar panels are now commonly used for grid connected power generation. In addition, this technology can be used to power cars, boats, houses, roadside telephones, remote sensors on gates, and various aspects businesses such as zoos and parks.

Since 2002, photovoltaic production has been increasing by an average of 20 percent a year. While over 90% of photovoltaic technology is used for grid-tied electric systems, an increasing number of off-grid uses are being discovered. Photovoltaic technology is the fastest growing energy technology in the world. It is estimated that by the year 2030, enough photovoltaic technology will exist to produce the solar power necessary to meet the electricity needs for 14 percent of the world's population.

Germany, Japan and the United States lead the world in photovoltaic installations, with those three countries representing 89 percent of the world's total installations. Germany also holds the record of being the fastest growing photovoltaic market in the world from 2006 to 2007, and by the end of 2006, almost 90 percent of all photovoltatic installations in Europe involved applications in Germany.

Materials for photovoltaic solar panels can vary. Silicon in various forms is one material that can be used. Some of these forms include monocrystalline, polycrystalline, and amorphous silicon. Crystalline silicon modules are estimated to account for up to 31 percent of all global installed power within the next two years. In addition, cadmium telluride and copper indium gallium sulfide can be used. Solar panels that are used as rooftop shingles and tiles or building facades are made to be flexible and thin, with semiconductor materials only being a few micrometers thick. In areas of the world with plenty of sun, a third generation of solar cells can be used. These cells can be made of solar inks created from dyes and conductive plastics. In addition, the new generation of solar cells utilizes plastic lenses or mirrors to concentrate sunlight onto small pieces of photovoltaic material.

Since all solar cells are only effective when used outdoors, solar cells require significant protection from environmental elements and hazards. Glass sheets are often used to protect tightly packed solar cells. Each solar panel typically holds 40 cells. The average home will require approximately 10 to 20 solar panels in order to generate enough electricity to power the house. Panels must be mounted on either a fixed angle facing south or on a device that allows the panels to track the sun. The mounted panels form an array, and hundreds of arrays can be connected to form a system used for industrial or electric plant applications.

With so many uses, it's easy to see why photovoltaic technology is quickly advancing and becoming a popular form of electricity production. As technology changes and advances and incentives to install systems are created, photovoltaic technology is fast becoming a viable option for consumers, from large-scale businesses to private residential homes.

New Words

photovoltaics [ˌfəutəuvɔl'teiiks] n.	光伏，太阳光电，太阳能光电板
electron [i'lektrɔn] n.	电子
electrode [i'lektrəud] n.	电极，电焊条
monocrystalline [ˌmɔnəu'kristəlain] n. & adj.	单晶体，单晶质；单晶的
polycrystalline [ˌpɔli'kristəlain] adj.	【晶体】多晶的
amorphous [ə'mɔːfəs] adj.	无定形的，无组织的，【物】非晶形的
crystalline ['krist(ə)lain] adj.	透明的，水晶般的，水晶制的
shingle ['ʃiŋgl] n.	瓦板
façade [fə'sɑːd] n.	（建筑物的）正面
lens [lenz] n.	透镜
array [ə'rei] n.	排列，阵列
viable ['vaiəbl] adj.	可行的
spacecraft ['speiskrɑːft] n.	宇宙飞船，航天器

Phrases and Expressions

solar radiation	太阳辐射，日光照射
solar panel	太阳能板，太阳能电池板
remote sensor	遥感器

cadmium telluride	碲化镉
copper indium gallium sulfide	铜铟镓硫化物

Notes

1. Photovoltaics (PV) is the field of technology and research related to the application of solar cells for energy by converting sunlight directly into electricity.

   译文：光伏（PV）是指以将太阳光直接转化为电能的太阳能电池为能源的应用技术和研究领域。

2. Photovoltaics is a green energy source that uses solar radiation in order to generate electric power.

   译文：光伏发电是一种绿色能源，它使用太阳辐射来发电。

3. During this process, the solar radiation excites electrons, which are then transferred between various bands, such as the conduction bands.

   译文：在此过程中，太阳辐射激发电子，然后这些电子将会在各种频带之间传送，如导电带。

4. However, due to an increase in the availability of photovoltaic technology as well as an increase in the demand for clean energy sources, solar panels are now commonly used for grid connected power generation.

   译文：然而，由于光伏技术有效性的增加，以及对清洁能源需求的增加，太阳能电池板现在常被运用到电网发电中。

   due to：因为，在句中可以作表语或状语，例如，(1) The team's success was largely due to her efforts. (2) The project had to be abandoned due to a lack of government funding.

   grid connected power generation：并网发电。

5. Crystalline silicon modules are estimated to account for up to 31 percent of all global installed power within the next two years.

   译文：未来两年内，在全球所有已安装的电源中，晶体硅模块估计会占到高达31%的比例。

   account for：（数量、比例上）占，例如，Computers account for 5% of the country's commercial electricity consumption.

6. It is estimated that by the year 2030, enough photovoltaic technology will exist to produce the solar power necessary to meet the electricity needs for 14 percent of the world's population.

   译文：据估计，到2030年将有足够的光伏发电技术利用太阳能发电以满足世界上14%的人口的电力需求。

Exercises

1. Write T (True) or F (False) beside the following statements about the text.

a. In photovoltaics, energy from the sun is converted into electrical current using conductors.

b. The solar radiation can excite electrons which are then transferred between various bands, such as the conduction bands.

c. Photovoltaic technology was first used to meet the demand for clean energy sources on the earth.

d. Since 2002, photovoltaic production has been increasing by an average of 90 percent a year.

e. Cadmium telluride and copper indium gallium sulfide cannot be used as materials for photovoltaic solar panels.

f. Solar panels should be mounted on a fixed angle facing south.

2. Fill in the missing words according to the text.

a. Photo refers to "_____" and voltaic to "_____".

b. Photovoltaic technology needs a _____, and primarily depends on _____.

c. Solar panels can be ground mounted, such as in _____ and in the form of _____.

d. Because of _____ in the availability of photovoltaic technology as well as an increase demand for _____, solar panels are now _____ for grid connected power generation.

e. As one material for photovoltaic solar panels, silicon can be used in various forms, such as _____, _____, and amorphous silicon.

f. The solar panels are mounted to _____ an array, and hundreds of arrays can _____ to form a system _____ industrial or electric plant applications.

3. Translate the following paragraph into Chinese.

Since all solar cells are only effective when used outdoors, solar cells require significant protection from environmental elements and hazards. Glass sheets are often used to protect tightly packed solar cells. Each solar panel typically holds 40 cells. The average home will require approximately 10 to 20 solar panels in order to generate enough electricity to power the house. Panels must be mounted on either a fixed angle facing south or on a device that allows the panels to track the sun. The mounted panels form an array, and hundreds of arrays can be connected to form a system used for industrial or electric plant applications.

Chinese Translation of Texts（参考译文）

第24课　光伏技术

photovoltaic 是一个组合词，photo 代表光，voltaic 代表电流。光伏（PV）是指以将太阳光直接转化为电能的太阳能电池为能源的应用技术和研究领域。

光伏发电是一种绿色能源，它使用太阳辐射来发电。它是把太阳辐射转化为可用电能的最广为人知的方法。在光伏电池中，来自太阳的能量通过半导体被转换成电流。这些半导体具有光伏效应。在此过程中，太阳辐射激发电子，然后这些电子将会在各种频带之间传送，如导电带，这将在电池的两个电极之间形成电压。光伏发电技术需要一个光源，它主要依赖来自太阳的能量。

光伏通常以太阳能电池板的形式出现。这些太阳能电池板里含有许多包含光电材料的太阳能

147

电池。这些面板可以在建筑物的屋顶上找到,包括家庭或企业的。此外,太阳能电池板可以安装在地面上,如在太阳能发电厂或者以光伏发电树的形式。

光伏技术的首次运用是为了给太空轨道上的卫星和航天器提供动力。然而,由于光伏技术有效性的增加,以及对清洁能源需求的增加,太阳能电池板现在常被运用到电网发电中。此外,这项技术可用于电力车、船、房屋、路边电话、门上的遥感器,以及各方面的业务,如动物园和公园。

自 2002 年以来,光伏产品一直在以平均每年 20%的幅度增加。虽然超过 90%的光伏技术用于并网电力系统,离网的用途也越来越多。光伏技术是世界上成长最快的能源技术。据估计,到 2030 年将有足够的光伏发电技术利用太阳能发电,以满足世界上 14%的人口的电力需求。

德国、日本和美国在光伏装置方面处于世界领先地位,这 3 个国家的光伏装置占世界总量的 89%。从 2006 年到 2007 年,德国也一直保持着其光伏市场在世界上发展最快的纪录,截至 2006 年底,欧洲近 90%的光伏装置都涉及德国的应用。

光伏太阳能电池板的材料也有所不同。硅是一种材料,有多种形式。这些形式包括单晶硅、多晶硅和非晶硅。未来两年内,在全球所有已安装的电源中,晶体硅模块估计占到高达 31%的比例。另外,也可以使用碲化镉和铜铟镓硫化物。用作屋顶瓦和砖或建筑外立面的太阳能电池板是灵活且薄得只有几微米厚的半导体材料。第三代太阳能电池可以使用在世界上某一个阳光充足的地方。这些电池可以由太阳能油墨构成,而这些油墨是由染料和导电塑料制成的。此外,新一代的太阳能电池利用塑料透镜或反射镜将太阳光聚集到小片的光伏材料上。

由于所有的太阳能电池只有在户外使用时才有效,所以太阳能电池需要有效的保护来抵御来自环境方面的危险和伤害。通常用玻璃板紧密包裹太阳能电池进行保护。每个太阳能电池板通常包含 40 个电池。平均每个家庭将需要大约 10 到 20 块太阳能电池板,以产生足够的电力给住宅供电。电池组件必须安装在一个固定的角度,朝向南方或者在可以让面板跟踪太阳的设备上。安装的面板形成一个阵列,数百个阵列可以连接成一个系统,用于工业或电力装置。

有了这么多的用途,我们很容易理解为什么光伏技术正在快速发展成为一种普遍的发电形式。随着技术的变化和发展,以及对安装光伏系统的鼓励和推广,从大型企业到私人住宅,光伏技术正迅速成为消费者的一种选择。

Reading Material

19. Digital Image Processing

Digital image processing generally refers to processing of a 2D picture using a digital computer (as shown in Fig.R19-1). In a broader sense, it implies digital processing of any 2D data. A digital image is an array of real or complex numbers represented by a finite number of bits. An image given in the form of a transparency, slide, photograph, or chart is first digitized and stored as a matrix of binary digits in computer memory—an RAM of great capacity. This digitized image can then be processed and/or displayed on a high-resolution monitor which refreshes at 3D frames/s to produce a visibly continuous display. Mini- or micro-computers are used to communicate and control all the digitization, storage, processing, and display operations via a computer network (such as the Ethernet). Program inputs to the computer are made through a terminal, and the outputs are available on a terminal, television monitor, or a printer.

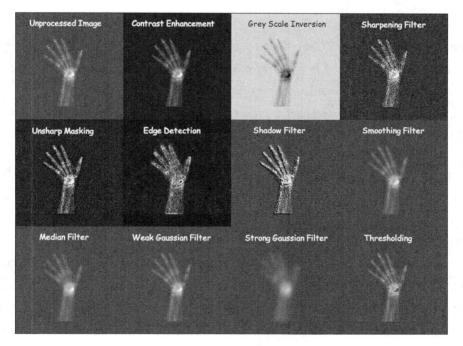

Fig.R19-1　Digital Image Processing

Digital image processing has a broad spectrum of applications, such as remote sensing via satellites and other spacecrafts, image transmission and storage for business applications, medicine, radar, sonar, and acoustic image processing, robotics, and automated inspection of industrial parts.

Images acquired by satellites are useful in tracking of earth resources; geographical mapping; prediction of agricultural crops, urban growth, and weather; flood and fire control; and many other environmental applications. Space image applications include recognition and analysis of objects contained in images obtained from deep space-probe missions. Image transmission and storage applications occur in broadcast television, teleconferencing, transmission of facsimile images (printed documents and graphics) for office automation, communication over computer networks, closed-circuit television-based security monitoring systems, and in military communications. In medical applications one is concerned with processing of chest X-rays, cineangiograms, projection images of transaxial tomography, and other medical images that occur in radiology, Nuclear Magnetic Resonance (NMR), and ultrasonic scanning. These images may be used for patient screening and monitoring or for detection of tumors or other disease in patients. Radar and sonar images are used for detection and recognition of various types of targets or in guidance and maneuvering of aircraft or missile systems. There are many other applications ranging from robot vision for industrial automation to image synthesis for cartoon making or fashion design. Whenever a human or a machine or any other entity receives data of two or more dimensions, an image is processed.

20. Computer Vision

Computer vision means artificial sight by means of computer and other pertinent techniques. Although today's computer vision systems are crude, compared with human sight, they are promising development of computer science and technology and will have brilliant prospects.

A typical real-time computer vision system includes the following components(as shown in Fig.R20-1).

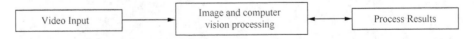

Fig. R20-1 Real-time Computer Vision System Components

Image Acquisition A TV camera is usually used to take instantaneous images and transform them into electrical signals, which will be further translated into binary numbers for the computer to handle. The TV camera scans one line at a time. Each line is further divided into hundreds of pixels. The whole frame is divided into hundreds (for example, 625) of lines. The brightness of a pixel can be represented by a binary number with certain bits, for example, 8 bits. The value of the binary number varies from 0 to 255 (2^8), a range great enough to accommodate all possible contrast levels of images taken from real scenes. These binary numbers are stored in an RAM (it must have a great capacity) ready for further processing by the computer.

Image Processing Image processing is for improving the quality of the images obtained. First, it is necessary to improve the signal-to-noise ratio. Here noise refers to any interference flaw or aberration that obscure the objects on the image. Second, it is possible to improve contrast, enhance sharpness of edges between images through various computational means.

Image Analysis　It is for outlining all possible objects that are included in the scene. A computer program checks through the binary visual information in store for it and identifies specific feature and characteristics of those objects. Edges or boundaries are identifiable because of the different brightness levels on either side of them. Using certain algorithms, the computer program can outline all possible boundaries of the objects in the scene. Image analysis also looks for textures and shadings between lines.

Image Comprehension　Image comprehension means understanding what is in a scene. Matching the prestored binary visual information with certain templates which represent specific objects in a binary form is a technique borrowed from Artificial Intelligence, commonly referred to as "template matching". One by one, the templates are checked against the binary information representing the scene. Once a match occurs, an object is identified. The template matching process continues until all possible objects in the scene have been identified, otherwise it fails (as shown in Fig.R20-2).

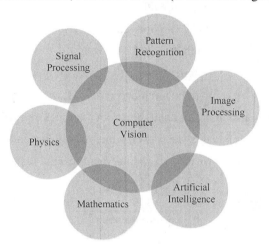

Fig. R20-2　Computer Vision

Computer vision has found its way into industries, doing jobs which used to be exclusive for human operators, for example, identifying specific objects or patterns by finding out distinctive details in shape, size, color, contrast, etc., inspecting products by checking for any flaws such as crack, smear. Jobs like these are boring and tiresome, and once the operator gets bored or tired, he/she tends to miss details or otherwise botch the manufacturing process.

One popular application of computer vision is machine vision. Giving vision to machines can automate their manufacturing process with little or no human intervention. Machine vision greatly improve productivity and quality, and the cost and time of manufacture can be dramatically reduced.

The promising application of computer vision is in robots. If a robot is equipped with computer vision, it becomes an intelligent robot. A robot with sight which would allow it to adjust its operations to fit itself to varying conditions and environments.

Today, computer vision is extensively used in non-industrial fields as well, for example, for identifying fingerprints or facial features of a suspect, distinguishing counterfeit notes and forged paintings, analyzing medical images and photographs taken by satellites, etc.

21. Multimedia Technology

Traditionally, computer systems dealt exclusively with numerical calculations. However, text processing soon became an important concern for computer designers. Communications technologies were also developed to support the transmission of textual and numerical data. More recently, there has been a dramatic increase in the range of media types supported by computers and communications technologies. Significant steps have been taken in integrating graphics into computer workstations and communications technology. Researchers are now tackling the harder problems presented by audio and video.

Two themes have emerged from this discussion. Firstly, the variety of media types is an important feature of modern information systems. Secondly, in order to deal with the variety, integration is a critical concern. These observations provide a good working definition of multimedia:

MULTIMEDIA = VARIETY + INTEGRATION

It is necessary for a multimedia system to support a variety of media types. This could be as modest as text and graphics or as rich as animation, audio and video. However, this alone is not sufficient for a multimedia environment. It is also important that the various sources of media types are integrated into a single system framework, a multimedia system is then one that allows end users to share, communicate and process a variety of forms of information in an integrated manner (as shown in Fig.R21-1).

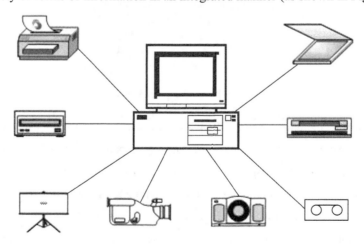

Fig.R21-1　Multimedia Computer System

Today, the various technologies referred to as Multimedia define a number of individual niches.

One of the most importance of these is animation, the capability to have moving images on your screen. Animation is tightly tied in with another concept called desktop video—actually creating and manipulating video images, to produce in-house presentations, rough drafts of commercial videos, or training products.

Sound will also play a key role in multimedia presentation.

Video images also take up a lot of disk space. To handle this, some groups are looking at optical discs for storage, particularly as erasable optical media became more mainstream.

Desktop video and animation is all well and good, but what many proponents see is a way of combining all these elements into an interactive system—interactive multimedia or hypermedia.

22. Artificial Neural Networks and Their Applications

Attempts to apply computer technology to image, speech, and forecasting share characteristics. All require pattern recognition, or the ability to identify or classify an entity despite noise and distortion. All hold great commercial promise because they reflect the human world. Finally, all are beginning to use neural networks to improve accuracy, reduce cost, or both.

A neural network is an implementation of an algorithm inspired by research into the brain (as shown in Fig.R22-1). In fact, one branch of neuroscience uses computers to model cognitive functions. But the neural networks discussed here have little to do with biology. Rather, they are a technology in which computers learn directly from data, thereby assisting in classification, function estimation, data compression, and similar tasks.

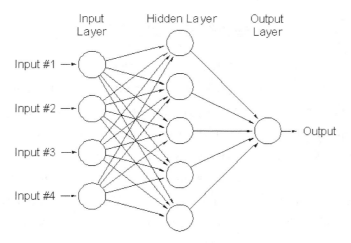

Fig.R22-1 Artificial Neural Networks

Having been used experimentally for decades, neural networks are reputedly a solution in search of a problem. More recently, though, they began moving into practical applications, and this trend can only accelerate now that specialized hardware is available to speed product development.

Hundreds of actual applications use the networks, often without public acknowledgment, to preserve a competitive advantage. Products known to be based on them include Optical Character Recognition (OCR) system; electrocardiograph; process control and financial systems. Military uses such as target recognition and flight control have also been reported, as well as communications applications like adaptive echo cancellation.

Computer scientists around the world are working to develop neural network software that mimics the human thought process more closely than digital computers do. Full-scale applications of the technology are years away, but they could include enhanced processing of sensory data, advanced artificial intelligence systems for decision support and robotics, among others.

Because the neural processes are embedded in silicon rather than software, orders-of-magnitude efficiency improvements over existing neural networks are possible.

Neural network programs are generally so large and complex that they require huge memory capacity and are a major processing challenge for even the most powerful computers available today.

Here are several ways that researchers are using neural network.

Pattern recognition Researchers commonly use filtering operations to solve image recognition and processing problems. These filtering operations are similar to the processing functions of neurons, so it was natural that the first major application for neural network was in imaging areas.

Nonlinear processes Neural networks also are useful for modeling nonlinear processes—those systems where the sum of the inputs is not directly proportional to the output.

Number crunching Computation-intensive applications also can be aided by neural network systems.

23. Radio Frequency Identification (RFID)

Radio Frequency Identification (RFID) is the wireless non-contact use of radio frequency electromagnetic fields to transfer data (as shown in Fig.R23-1), for the purposes of automatically identifying and tracking tags attached to objects. The tags contain electronically stored information. Some tags are powered by and read at short ranges (a few meters) via magnetic fields (electromagnetic induction). Others use a local power source such as a battery, or else have no battery but collect energy from the interrogating EM field, and then act as a passive transponder to emit microwaves or UHF radio waves (i.e., electromagnetic radiation at high frequencies). Battery powered tags may operate at hundreds of meters. Unlike a barcode, the tag does not necessarily need to be within line of sight of the reader, and may be embedded in the tracked object.

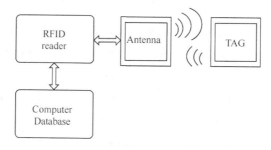

Fig.R23-1 A Radio Frequency Identification System

RFID tags are used in many industries. An RFID tag attached to an automobile during production can be used to track its progress through the assembly line. Pharmaceuticals can be tracked through warehouses. Livestock and pets may have tags injected, allowing positive identification of the animal. On off-shore oil and gas platforms, RFID tags are worn by personnel as a safety measure, allowing them to be located 24 hours a day and to be quickly found in emergencies.

Since RFID tags can be attached to clothing, possessions, or even implanted within people, the possibility of reading personally-linked information without consent has raised privacy concerns.

A Radio Frequency Identification system uses tags, or labels attached to the objects to be identified. Two-way radio transmitter-receivers called interrogators or readers send a signal to the tag and read its response.

RFID tags can be either passive, active or battery-assisted passive. An active tag has an on-board battery and periodically transmits its ID signal. A Battery-Assisted Passive (BAP) has a small battery on board and is activated when in the presence of a RFID reader. A passive tag is cheaper and smaller because it has no battery. However, to start operation of passive tags, they must be illuminated with a power level roughly three magnitudes stronger than for signal transmission. That makes a difference in interference and in exposure to radiation.

Tags may either be read-only, having a factory-assigned serial number that is used as a key into a database, or may be read/write, where object-specific data can be written into the tag by the system user. Field programmable tags may be write-once, read-multiple; "blank" tags may be written with an electronic product code by the user. A tag with no inherent identity is always threatened to get manipulated.

RFID tags contain at least two parts: an integrated circuit for storing and processing information, modulating and demodulating a Radio Frequency (RF) signal, collecting DC power from the incident reader signal, and other specialized functions; and an antenna for receiving and transmitting the signal. The tag information is stored in a non-volatile memory. The RFID tag includes either a chip-wired logic or a programmed or programmable data processor for processing the transmission and sensor data, respectively.

An RFID reader transmits an encoded radio signal to interrogate the tag. The RFID tag receives the message and then responds with its identification and other information. This may be only a unique tag serial number, or may be product-related information such as a stock number, lot or batch number, production date, or other specific information.

RFID systems can be classified by the type of tag and reader. A Passive Reader Active Tag (PRAT) system has a passive reader which only receives radio signals from active tags (battery operated, transmit only); as shown in Fig.R23-2. The reception range of a PRAT system reader can be adjusted from 1–2000 feet (0.30–610m), allowing flexibility in applications such as asset protection and supervision.

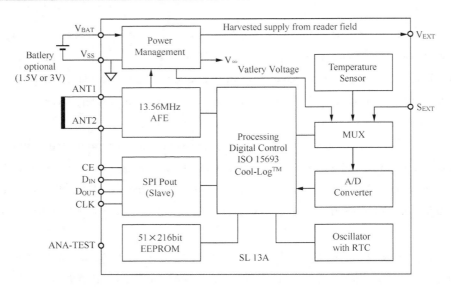

Fig.R23-2　Passive Reader Active Tag (PRAT) System

24. Intelligent City

Intelligent city centers on urbanization development, sustainable city growth and core demands of urban residents. Intelligent city requires effective integration of advanced information technology with advanced operating and service philosophy (as shown in Fig.R24-1). An intelligent city will collect and store a multitude of a city's information resources in real time to create its IT infrastructure, and by data interconnection and interoperability, exchange and sharing, and collaborated applications, it will create a platform which provides a convenient, efficient, and flexible tool for generating and implementing decisions related to the city management and operation, as well as for the provisioning and management of innovative public services, with an ultimate goal of achieving harmonious development of safer, greener, more efficient and more convenient urbanization.

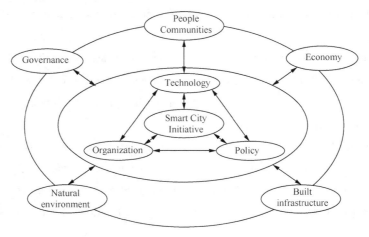

Fig.R24-1　Intelligent City

Every smart city should have is presented with examples of applications of these in different cities. The list of the technologies is the following.

1. Open-data initiatives and hackathons
2. Parking apps
3. Apps that let users "adopt" city property
4. High-tech waste management systems
5. All-digital and easy-to-use parking payment systems
6. A city guide app
7. Touchscreens around the city
8. Wi-Fi in subway stations and on trains
9. Sustainable and energy efficient residential and commercial real estate
10. Dynamic kiosks that display real-time information
11. App or social media-based emergency alert and crisis response systems
12. Police forces that use real-time data to monitor and prevent crime
13. More public transit, high-speed trains, and Bus Rapid Transit (BRT)
14. OLED lights and surveillance in high-crime zones
15. Charging stations, like the solar-powered
16. Roofs covered with solar panels or gardens
17. Bike-sharing programs
18. A sharing economy, instead of a buying economy
19. Smart climate control systems in homes and businesses
20. Widespread use of traffic rerouting apps
21. Water-recycling systems
22. Crowdsourced urban planning
23. Broadband Internet access for all citizens
24. Mobile payments
25. Ride-sharing programs

Unit V

Basic Knowledge of Professional English Ⅰ

专业英语基础知识 Ⅰ

> 本单元介绍了产品说明书阅读（翻译）的特点和技巧，单元中还列举了不同样式的产品说明书，有条款（条文）式产品说明书、图表式产品说明书、图文式产品说明书、条款（条文）和图表结合式产品说明书等。通过本单元的学习，学生可以了解产品说明书翻译的特点和技巧，提高专业英语的阅读（翻译）与写作水平。

产品说明书阅读（翻译）的特点和技巧

5.1 概述

产品说明书是科技文体的一种，它以传递产品的有用信息为主要目的，内容主要包括前言、部件、基本功能、使用指南，故障排除等。其中，基本功能和使用指南是主体部分。

一些简单易用的电器、电子产品，其说明书也相对简略。

产品使用说明书有两类：一类是包装式，即印在包装物上的说明书；另一类是内装式，它们均装在包装物内。

产品使用说明书样式多样，有单页、活页、卡片、小册子。说明书形式有条款（条文）式产品说明书、图表式产品说明书、图文式产品说明书、条款（条文）和图表结合式产品说明书等。

5.2 产品说明书的特点

特点概括起来包括准确性（Accuracy）、简明性（Conciseness）、客观性（Objectivity）等。

5.2.1 准确性

产品说明书是为了指导读者正确使用产品而写的，它传递的信息（如各种数据、图表）

必须科学准确。在翻译过程中，必须把信息内容如实准确地翻译出来，显化原文隐含的信息，消除歧义。一些专业术语、固定用语和习惯说法必须表达得准确、地道，例如，在翻译数码相机说明书时会遇到这样一些术语：镜头后盖（Ear Lens Cap）、三脚架（Tripod）、数码变焦（Digital Zoom）、快门帘幕（Shutter Curtain）、曝光不足（Under Exposure）、取景器（View Finder）等，这些都需要按专业说法表达出来，不可任意生造。

5.2.2 简明性

简明性的特点表现为以下几点。

（1）内容条目简洁明了，步骤清晰，逻辑性强。

（2）常用缩略形式，如下所示。

- 液晶显示（Liquid Crystal Display）常缩写成 LCD。
- 发光二极管（Light Emitting Diode）常缩写成 LED。
- 中央处理器（Central Processing Unit）常缩写成 CPU。
- 自动对焦（Auto Focus）常缩写成 AF。
- 手动对焦（Manual Focus）常缩写成 MF。

5.2.3 客观性

产品说明书可将该产品的相关内容客观地呈现出来，引导读者按照一定的思维逻辑循序渐进，使读者知道该做什么，怎么做，进而了解和正确使用该产品。这些内容带有描述说明的性质，客观而不带有感情色彩。

例如：The emergency-eject option allows the user to naturally open the CD tray during a power malfunction.

译文：紧急退出功能键可让使用者在电源故障时，以手动方式打开 CD 托盘。

5.2.4 准确性、简明性、客观性的共同体现

产品说明书的翻译具有准确、简明、客观等特点，这些特点共同体现在以下方面。

（1）广泛使用复合名词结构，以求行文简洁、明了、客观。

例如：equipment check list。

译文：设备清单。

例如：warranty card。

译文：保修卡。

译句常使用非人称名词化结构作主语，使句意更客观、简洁。

例如：The use of computers has solved the problems in the area of calculating.

译文：由于使用了计算机，数据计算方面的问题得到了解决。

（2）普遍使用一般现在时。一般现在时可以用来表示不受时限的客观存在，包括客观真理、格言、科学事实及其他不受时限的事实。

产品说明书的主体部分就是进行"无时间性"（Timeless）的一般叙述，其译文普遍使用一般现在时，以体现出内容的客观性和形式的简明性。

例如：This facsimile machine is not compatible with digital telephone systems.

译文：本传真机与数码电话系统不兼容。

（3）常使用被动语态。产品说明书英译的主要目的是说明相关产品（即受动者）的客观事实，其强调的是所叙述的事物本身，而并不需要过多地注意它的行为主体（即施动者）。

例如：Additional information can be found in the electronic user's manual which is located on the CD-ROM.

译文：您可以在光盘的电子使用手册中找到额外的信息。

（4）广泛使用祈使句。产品说明书很多地方都是指导使用者要做什么，不要做什么，或该怎么做，所以经常使用祈使句，谓语一般用动词原形，没有主语，译文的表述显得准确、客观而又简洁、明了。

例如：Do not store CF cards in hot, dusty or humid places. Also avoid places prone to generate static charge or an electromagnetic field.（文中出现了两个祈使句）

译文：请勿将 CF 卡存放在过热、多灰尘，或潮湿的环境中，也不能将其存放在能产生静电荷或者电磁波的环境中。

再看 PHILIPS 显示器安装指南的译文，如下所示。

（1）Turn off your computer and unplug its power cable.

（2）Connect the blue connector of the video cable to the blue video connector on the back.

（3）Connect your monitor's power cable to the power port on the back of the monitor.

（4）Plug your computer's power cord and your monitor into a nearby outlet.

（5）Turn on your computer and monitor. If the monitor displays an image, installation is complete.

（6）If you are using BNC connectors (not available on all models), please remember to switch to "Input B" from "Input A" by simultaneously pressing the "OK" and "UP" knobs on front control panel.

以上 6 句都是祈使句，可见祈使句在产品说明书英译时使用的广泛性。

5.3　产品说明书的英译技巧

产品说明书的很大篇幅是在叙述使用方法和操作步骤，其语言平实，修辞手法单调，很少用到文学作品中常出现的比喻、拟人、夸张等修辞手法。因此，其译文也比较平实，翻译时以直译为主。

5.3.1　直译（Literal Translation）

在将原文的产品说明书翻译成译语时，直译是最常用的技巧。

例如：The battery's service life is 10 years.

译文：电池的使用寿命是 10 年。

5.3.2　意译（Free Translation）

在翻译过程中，将原文的一些词语或句子成分做适当调整，才能使译文更好地符合译语的表达习惯，这时就需要运用意译这一重要的翻译技巧，具体包括以下几种。

（1）语序调整。汉英两种语言有不同的表达习惯，词和分句的顺序有时也不一样，如在表示时间地点时，汉语习惯先大后小，而英语则习惯先小后大。

例如：No pictures on the monitor.

译文：监视器上没有影像。

以下是深圳某电子公司的地址，在翻译时，其词语顺序也做了调整。

例如：Room308, Building 402, #52 Zhenhua Road, Futian, Shenzhen, China

译文：中国深圳市福田区振华路 52 号 402 栋大厦 308 室

（2）词类转换。翻译不是机械照搬，在做翻译时，原文的某些词类应根据译语的表达习惯做适当转换才能使译文更自然、地道。动词、名词、形容词、副词等在翻译时都能转换成其他词类。

例如：Some people are for the operation method, while some are against it.

译文：有些人赞成这个操作方法，而有些人反对。

for 和 against 两个介词翻译成了动词"赞成""反对"。

例如：This machine helps the blind to walk。

译文：这个机器帮助盲人行走。

形容词 blind 翻译成了名词"盲人"。

例如：You'll find this sketch map of great use。

译文：你会发现，这份示意图是十分有用的。

use 翻译成了"有用的"。

5.4 训练

（1）教材提供了几个产品说明书（节选），请自行阅读翻译，再对照参考译文以检查学习效果。

（2）试着写出一款电子产品的英文说明书，注意说明书的编写要求。

5.4.1 产品说明书资料 1

Portable Multimedia Acoustics

User's Manual

Concentrate on perfect sound pursuit…

Thank you for using this digital product of our company. In order to let you experience the product swimmingly, detailed instruction is provided which you can find the product's introduction, usage and other information. Before using this product, please read the manual carefully, so that you can correctly use it. In case of any printing or translation error, we apologize for the inconvenience. As for the content change, we are sorry for no further notice.

1. General Information

This product is a well-designed portable multimedia mini acoustics which applies to household, outdoor travel, office and other places. It offers you a chance to indulge yourself in music at any time and place, and provides perfect sound service to your computers, digital media players, mobile phones and other audio-visual products.

2. Function Overview

[MP3 Player] Enable directly playing MP3 files in TF memory card and U Disk.

[FM Radio] A FM digital stereo radio; enable mnemonic radio channel playing.

[Ext Earphone] A standard earphone jack included; enable listening with earphone.

[Audio input] A stereo audio input jack; enable the sound source connection with computers, digital media players, mobile phones and other audio-visual products.

[Memory Function] Memorize the item played last time automatically.

[Chargeable Battery] Chargeable battery included; environmental, economical and functional.

[Intellective charging] Wiring with USB jack, enables charging by connecting with USB jack of computer or cell phone charger.

3. Operation

Microcomputer system of this product can automatically identify exterior equipment. After startup and entering into standby mode, insert U disk, TF card or audio signal lines for automatic identification and being played according to the principle of the later coming first. For more details, please reference to 4th item "Definition of Button and Jack".

4. Definition of Button and Jack（Real object be taken as final）

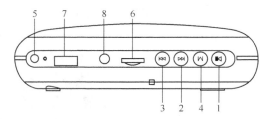

(1)【○/ ▶Ⅱ】：Long press for Power On/Off, short press for Play/Pause. In FM mold, protected radio channel can be chosen.

(2)【◀◀/V−】：Short press for last item or last channel, long press for turning down the volume.

(3)【V+/▶▶】：Short press for next item or next channel, long press for turning up the volume.

(4)【MODE】：Mode shift button: short press and shift to USB/SD mode, once more to LINE IN (AUX audio input mode); in FM mode, long press stands for automatic search and memorizing the radio channel.

(5)【DC 5V/Ω】：Power input + earphone jack for listening to music or radio channel. Power input jack with USB wiring, enables charging by connecting with USB jack of computer or cell phone charger.

(6)【TF CARD】：Insert TF card for MP3 music playing.

(7)【USB】：Insert U disk for MP3 music playing.

(8)【AUX】：Audio input jack + exterior antenna jack, enable the sound source connection with computers, digital media players, mobile phones and other audio-visual products.

5. Introduction of Indicator Light

Red: FM radio mode.

Blue: U disk or TF card playing mode.

Purple: audio input mode.

Twinkle of purple indicator light: under charging; no twinkle after fully charged.

6. Function and Usage Method of Radio（Optional type, Real object be taken as final）

(1) In FM mode, press "◄◄" or "►►" to choose "last channel" or "next channel".

(2) Volume adjusting, long press "◄◄" to turn down, press "►►" to turn up.

(3) Automatic search: in FM mode, long press "MODE" for automatic search; director light twinkling when in searching; channel be memorized and saved after search finished.

(4) Memorized channel choosing: after search finished, pressing " ►I " can choose saved channel circularly.

(5) Audio input wiring inserted into "AUX" can be used as radio antenna; in this way better effect can be received. As shown in fig.

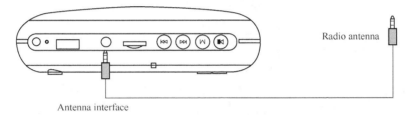

Antenna interface

(6) Search with earphone, better effect can be received.

7. Charging Method

One end of wiring be inserted in DC 5V, another end with USB jack be inserted in USB jack of PC or other charging with standard 5V 500mA jack; twinkle under charging and no twinkle after fully charged.

8. Low Power

When low power occurs and system been power off automatically, please charge in time; not use this product for a long period, please charge at least once a month so as to lengthen battery's life.

9. FQA

Power off automatically even after startup: low power, please use after 2 hours' charging.

Power off automatically or restart: low power, please use after 2 hours' charging.

Music cannot be played: MP3 files' saving path cannot be identified by player; please save files into disk's root folder.

No sound: check whether the volume has been turned on.

No sound after connecting with computer: incorrect connection, please connect correctly.

No response of button or function error: restart or pull out battery then put in again.

10. Technical Specification

Output Power: 3 W(RMS)

Frequency Effect: 100 Hz~18 kHz

Input Power: direct current DC5V 500 mA

Maximum Power Consumption: 400 mA

Audio signal input jack: standard ϕ 3.5 mm jack of stereo earphone.

11. Accessories

(1) a main body

(2) a set of User's Manual & Warranty card

(3) a USB wiring

(4) a audio input wiring

（选自某公司产品说明书）

Chinese Translation of Texts（参考译文）

便携式多媒体音响

使用手册

专注于对完美音质的追求……

感谢您使用本公司出品的数码产品，为了让您轻松体验产品，我们随机配备了内容详尽的使用说明，您从中可以获取有关产品的介绍、使用方法等方面的知识。在您开始使用本机之前请先仔细阅读说明书，以便您能正确地使用本机，如有任何印刷错误或翻译失误望广大用户谅解，当涉及内容有所更改时，恕不另行通知。

1. 产品概述

本机是一款外观小巧、设计精美、携带方便的多媒体小音响，适用于家居、户外旅游、办公室等场所，它能让您随时随地享受音乐带来的轻松，为您的计算机、数码音乐播放器、手机等视听产品提供完美的音质。

2. 功能特点

【MP3 播放】直接播放 TF 卡及 U 盘 MP3 文件。

【FM 收音机】FM 数字立体声收音机，电台记忆播放。

【外挂耳机】本机配置了标准耳机插孔，您可选配耳机聆听音乐。

【音频输入】立体声音频输入接口，轻松接驳计算机、数码音乐播放器、手机等视听产品等各类音源。

【断点记忆】自动记忆上次退出时的曲目播放。

【可充锂电】内置可充电锂电池，环保、节约、实用。

【智能充电】配送 USB 接口充电线，可接驳计算机的 USB 接口进行充电，或使用手机充电器进行充电。

3. 播放音乐操作

本机的微型电子计算机系统自动检测识别外接设备，开机后进入待机模式，插入 U 盘或 TF 储存卡后可自动识别播放，插入音频信号线后也可自动识别播放，其遵循后者优先原则，详细的功能操作请阅读第 4 项"产品的按键、插孔功能定义"。

Unit Ⅴ　Basic Knowledge of Professional English Ⅰ　专业英语基础知识Ⅰ

4. 产品的按键、插孔功能定义（以实物为准）

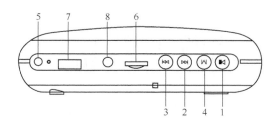

（1）【○/ ▶Ⅱ】：长按开机/关机，播放/暂停。在 FM 模式下可以选择已被保存的电台。

（2）【◀◀/V−】：短按上一曲，选择上一个收音电台，长按调节音量减小。

（3）【V+/▶▶】：短按下一曲，选择下一个收音电台，长按调节音量增大。

（4）【MODE】：模式转换键，短按转为 USB/SD 卡模式，再短按为 LINE IN(AUX 音频输入模式)，在 FM 收音模式下，长按为自动搜台并记忆保存电台。

（5）【DC 5V/Ω】：电源输入+耳机公用插孔，可以插耳机欣赏音乐或收听电台；电源输入插孔，可使用本机配送的专用 USB 电源线，接驳计算机 USB 接口进行充电，或使用手机充电器进行充电。

（6）【TF CARD】：插入 TF 卡播放 MP3 音乐。

（7）【USB】：插入 U 盘播放 MP3 音乐。

（8）【AUX】：音频输入接口+外置天线插孔，接驳计算机、数码音乐播放器、手机等视听产品等各类音源。使用收音机功能时，参照第 6 项中的图示连接外置天线，收台灵敏度更佳。

5. 指示灯介绍

红色指示灯：FM 收音机模式。

蓝色指示灯：U 盘和 TF 存储卡播放模式。

紫色指示灯：音频输入模式。

紫色指示灯闪动：正在充电中，充满电量后停止闪动。

6. 收音机功能使用方法（可选机型，功能以实物为准）

（1）在 FM 收音模式下，按"◀◀"或"▶▶"键可选择"上一个电台"或"下一个电台"。

（2）音量调节，长按"◀◀"键减小音量，长按"▶▶"键增大音量。

（3）自动搜台：FM 收音状态，长按"MODE"键开始自动搜台。搜台时指示灯闪烁，搜台完毕后自动记忆保存电台。

（4）记忆选台：自动搜台完毕后，按"▶Ⅱ"键可循环选择已被保存的电台频道。

（5）将"音频线"插到"AUX"插孔可做本机收音天线，收台效果更佳，如图所示。

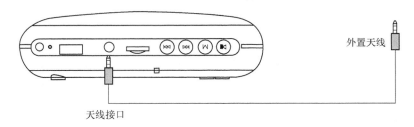

165

（6）戴耳机收听电台，收台效果也会提高。

7. 充电使用方法

将电源线一端插入本机的 DC5V 插孔，另一端 USB 插头插入 PC 的 USB 接口，或连接标准 5V 500mA 的充电器接口，充电时"紫色灯"会闪动，充满电量后"紫色灯"会停止闪动。

8. 电池低电

当电量出现低电压时，系统会自动关机，此时请及时充电；如果长时间不使用本机，至少每月充电一次，可保护电池，延长电池寿命。

9. 疑难解答

开机就自动关机：电池电量不足，请充电两小时后再使用。

播放自动关机或重启：电池电量不足，请充电两小时后再使用。

不能播放音乐：MP3 音乐文件的存放路径不能被播放器识别，请将文件存放在可移动设备的根目录下。

喇叭无声音：音量是否打开。

连接计算机无声音：计算机接口连接错误，请选择正确的计算机音频输入接口。

按键无功能或功能错：关机后再开机，或拨下电池再重新装上。

10. 技术规格

输出功率：3 W(RMS)

频率效应：100 Hz~18 kHz

输入电源：直流 DC5V　500mA

最大消耗电源：400 mA

音频信号输入接口：标准 ϕ3.5 mm 立体声耳机插孔。

11. 包装附件

（1）主机一台

（2）使用说明书、保修卡一套

（3）USB 电源线一条

（4）音频转接线（计算机信号线）一条

5.4.2　产品说明书资料2

High-definition camera pen manual

Picture for reference only, please

1．**The Structure Diagram**

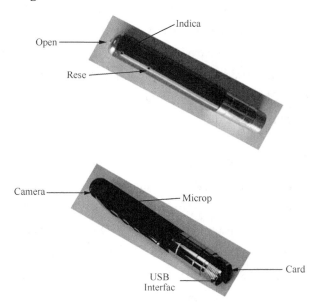

Picture for reference only, please

2．**Introduction**

This product is a high-definition audio and video capture function of a digital pen that can capture and store high-definition video with sound. Is a simple, small and exquisite, beautiful and practical, easy to carry features for business, education, security, media, justice, tourism, health, life and other areas of essential utilities, deeply favored customers.

3. Operation Instructions

(1) Switch Machine: long press the button two seconds to complete the boot into standby mode, this time a long green indicator light cameras; in a light lit in any state, long press to open the key, the camera will automatically save the current photo recorded in the file and shut down.

(2) Recording Video: camera into the camera to wait for the state, this time a long green light, long press the button, the green light flashes three times to start video; short press the button again to stop recording, save the file, this time to restore a long bright red and green; in order to ensure shot file size and security, the system will automatically save every 50 minutes, once a file, and continue recording.

(3) Photo: boot, after the camera into the camera to wait for the state, this time a long green light, short press the button, the green light flashes about to start taking pictures and save; press time, the green light flashes about to start taking pictures and save; cycle.

(4) PC Camera: boot in any mode of state, to connect a computer USB port, you can enter the PC Camera (webcam) mode, in this mode, you can use the camera as a PC camera, with your friends online video chat, shooting emoticons, and so on.

(5) Connected to a computer: in the off state or standby state, can be connected to a computer at this time, the computer will pop up a removable disk logo, image file directory as follows: * Plate \ _REC \ 100MEDIA; when the camera and computer to exchange data, the green light will blink rapidly; need to uninstall the camera, please click on the task bar bottom right corner of the mobile device management icons inside uninstall the removable storage device, to be systematically identified. After the hardware can be safely pulled out, then disconnect the camera and computer connections.

(6) Charging: connecting a computer USB port or charger, can charge the camera, when the battery is charging state, the red light will blink slowly when the battery is fully charged, the red light into a long-Liang.

(7) Set the time: the camera provides a video file to display the time recording function, you may need to set the time according to the actual value; time display format: "year. Dated. Day: minutes: seconds", set the time as follows.

① Turn off the camera → Connect the computer → Open the camera removable disk → In the space below the root directory of the camera click the right mouse button → to move the mouse pop-up menu "New" option at the top → Select "Text Document" → to the text document named "time", to note its extension ". Txt" → time to set file built; you can follow the steps above on your desktop, other disks Division, completed the other folder and then copied to the removable disk root directory.

② Open the "time" a text document, enter the time settings inside the string, the string by "date when the minutes and seconds", composed of several parts, to set the time, to set the following format:

2009.01.01 12.01.01 Y

The format string portion of the time you want to set time value, pay attention to date in the spaces with the minutes and seconds.

③ To confirm time setting file has been copied to the root directory, uninstall the hardware, boot, time is set to complete.

④ Setting a good time will be saved to the camera inside.

Exception handling

Camera storage space is insufficient, while in the green and red light flashes about 5 seconds later, automatically save the current image file and shut down. If you want to continue to use the camera, in the image on the computer back up important files, and delete the memory of the old files to free up enough storage space.

Camera battery is insufficient, margin will be alternately flashes green and red light about 5 seconds later, automatically save the current image file and shut down. If you want to continue to use the camera, please charge.

The camera due to accidental causes improper operation or other special stops responding, you can use toothpicks and other non-metallic thin rod stretching reset hole to reset.

4. The Relevant Parameters

Item	Parameters
Video Format	AVI
Video Coding	M-JPEG
Video Resolution	1280*960 VGA
Video Frame Rate	30±1fps
Player	Or mainstream operating system comes with audio and video playback software
Image Ratio	4:3
Support System	Windows Me/2000/XP/2003/Vista
Charge Voltage	DC 5V
Interface	Mini USB interface
Storage Support	TF Card
Battery Type	High-capacity polymer lithium battery

5. Notes

(1) Use the occasion: please strictly abide by relevant state laws, this product can not be used for any unlawful purpose, or peril.

(2) With regard to the battery: with the growth in use of time, the battery will be shortened working hours. Long-term is not used, please fully charged before use.

(3) File security: this product is non-professional storage devices, internal storage of documents does not guarantee the integrity and security, real-time on the computer or other storage device to back up your important documents.

(4) Video quality: this product is non-professional video recording equipment, do not guarantee the effectiveness of video files to achieve your expectations.

(5) Working Temperature : 0℃~40℃.

(6) Operating humidity: 20%~80%, do not place wet product work environment, the product does not have the waterproof function.

(7) Shoot illumination: do not facing the camera directly to the sun and other strong light source, so as to avoid optical devices damaged.

(8) Cleaning request: do not dust density in the environment is too large to use in order to avoid dust contamination of lenses and other components affect the camera effects. Lens glasses with lens cleaning paper or cloth gently wipe to keep clean.

(9) Other matters: the product belongs to sophisticated electronic products, please do not make it have a strong impact, vibration; do not in strong magnetic fields, strong electric field to use.

(10) Supplementary Note: other non-specified matters please contact your local dealer.

（选自某公司产品说明书）

Chinese Translation of Texts（参考译文）

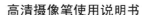

高清摄像笔使用说明书

图片仅供参考，请以实物为准

1. 结构示意图

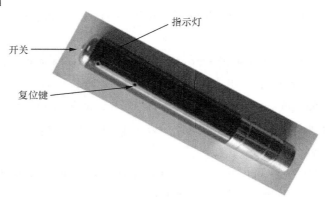

Unit Ⅴ　Basic Knowledge of Professional English Ⅰ 专业英语基础知识 Ⅰ

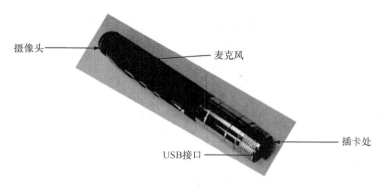

图片仅供参考，请以实物为准

2. 产品简介

本产品是具有高清影音拍摄功能的数码笔，可以拍摄和存储高画质有声视频。本产品具有操作简便、小巧精致、美观实用、便于携带的特点，是商务、教育、安防、媒体、司法、旅游、医疗、生活等领域必备的实用工具，深受广大用户青睐。

3. 操作说明

（1）开关机：长按按钮 2 秒钟，完成开机，进入待机状态，此时摄像机指示灯绿灯长亮；在有指示灯点亮的任意状态下，长按开关键，摄像机将自动保存当前的摄录文件并关机。

（2）录制视频：摄像机进入摄像等待状态，此时绿灯长亮，长按按钮，绿灯闪烁三下，开始录像；再次短按按钮，停止录像，保存文件，此时红绿恢复长亮；为了保证拍摄文件的大小和安全，系统将每隔 50 分钟自动保存一次文件，并继续录像。

（3）拍照：开机后，摄像机进入摄像等待状态，此时绿灯长亮，短按按钮，绿灯闪烁一下，开始拍照并保存；再按一次，绿灯闪烁一下，开始拍照并保存；周而复始。

（4）PC 摄像头：在任意模式的开机状态下，连接计算机的 USB 接口，即可进入 PC Camera（网络摄像头）模式，在该模式下，你可以将摄像机作为计算机摄像头使用，与好友进行网上视频聊天、拍摄大头贴，等等。

（5）连接计算机：在关机状态或者待机状态时，可以连接计算机，此时，计算机将会弹出可移动磁盘标识，影像文件保存目录为*盘\ _REC\100MEDIA；当摄像机与计算机交换数据时，绿灯会快速闪烁；需要卸载摄像机时，请单击任务栏右下角的可移动设备管理图标，在里面卸载该可移动存储设备，待系统确认可以安全拔出硬件后，再断开摄像机与计算机的连接。

（6）充电：连接计算机的 USB 接口或充电器，即可对摄像机进行充电，当电池处于充电状态时，红灯将慢速闪烁，当电池充满电后，红灯变为长亮。

（7）设置时间：摄像机提供了在视频文件中显示录像时间的功能，你可以根据实际需要设置时间的值；时间显示格式为"年.月.日 时:分:秒"，设置时间的方法如下。

① 关闭摄像机 → 连接计算机 → 打开摄像机可移动磁盘 → 在摄像机根目录下面的空白处单击鼠标右键 → 把鼠标移至弹出菜单的"新建"选项上方 → 选择"文本文档" → 给该文本文档取名为"time"，需要注意它的扩展名为".txt" → 时间设置文件建成；你也可以按照上述步骤在计算机桌面、其他磁盘分区、其他文件夹下建好后再复制到可移动磁盘的根目录下。

② 打开"time"文本文档，在里面输入时间设置字符串，字符串由"年月日 时分秒"几个部分组成，要设置时间，请按以下格式设置。

2009.01.01 12.01.01 Y

上述格式中的时间字符串部分为你想设置的时间值，注意年月日与时分秒中的空格。

③ 确认时间设置文件已经复制到根目录下之后，卸载硬件，开机，时间设置完成。

④ 设置好的时间将保存到摄像机里面。

异常处理情况如下。

摄像机存储空间不足时，将在绿灯和红灯同时闪烁约 5 秒后，自动保存当前影像文件并关机。如果你要继续使用摄像机，请在计算机上备份重要的影像文件，并删除存储器内的旧文件，以腾出足够的存储空间。

摄像机电池余量不足时，将在绿灯和红灯交替闪烁约 5 秒后，自动保存当前影像文件并关机。如果你要继续使用摄像机，请先充电。

摄像机因偶然的不当操作或其他特殊原因停止响应时，可以用牙签等非金属细棍伸入复位孔进行复位。

4. 相关参数

项　　目	相关参数
视频格式	AVI
视频编码	M-JPEG
视频分辨率	1280*960 VGA
视频帧率	30±1fps
播放软件	操作系统自带或主流影音播放软件
影像比例	4∶3
支持系统	Windows Me/2000/XP/2003/Vista
充电电压	DC 5V
接口类型	Mini USB 接口
存储支持	TF 卡
电池类型	高容量聚合物锂电

5. 注意事项

（1）使用场合：请严格遵守国家相关法令，不得将此产品用于任何非法用途，否则后果自负。

（2）关于电池：随着使用时间的增长，电池工作时间会有所缩短。长久未使用，请在使用前先充满电。

（3）文件安全：本产品是非专业存储设备，不保证内部存储文件的完整性和安全性，请即时在计算机或者其他存储设备上备份您的重要文件。

（4）摄录品质：本产品是非专业摄录设备，不保证所摄录文件的效果能达到您的期望。

（5）工作温度：0℃～40℃。

（6）工作湿度：20%～80%，请勿将产品置于潮湿的工作环境，产品不具备防水功能。

（7）拍摄照度：请勿将摄像头直接对着太阳等强光源，以免光学器件受到损伤。

（8）保洁要求：请勿在粉尘密度过大的环境下使用，以免镜头及其他部件沾染粉尘，影响摄像效果。镜头可以用擦镜纸或眼镜布轻轻擦拭，保持洁净。

（9）其他事项：该产品属于精密电子产品，请勿使其受到强烈冲击、震动；请勿在强磁场、强电场下使用。

（10）补充说明：其他未明事宜请与当地经销商联系。

5.4.3　产品说明书资料3

Instructions

BeoPlay A1 User Guide

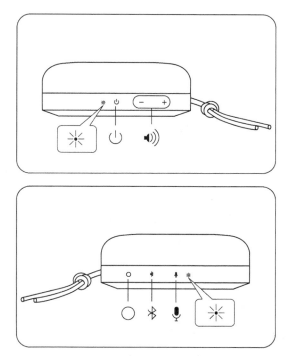

Charging

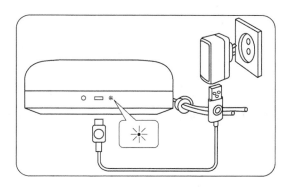

Charging time is approximately 3 hours using the included USB cable.

Battery Indicator…

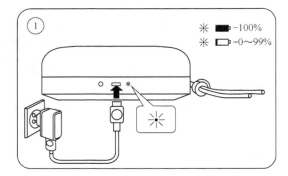

The battery indicator flashes orange when charging.

The battery indicator turns green when the battery is fully charged.

…Battery Indicator

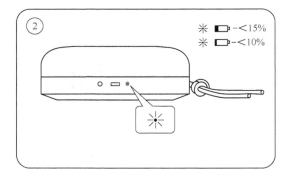

When the battery level is below 15%, the battery indicator turns solid red. When the battery level is below 10%, the battery indicator starts flashing red.

Bluetooth pairing…

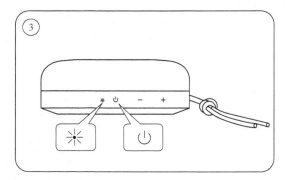

Press the power button to switch BeoPlay A1 on. The product indicator turns solid white, and the BeoPlay A1 is ready to be set up.

Unit Ⅴ Basic Knowledge of Professional English Ⅰ 专业英语基础知识 Ⅰ

Bluetooth pairing...

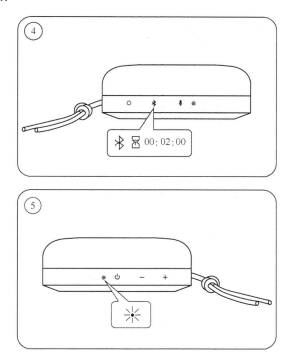

Press and hold the Bluetooth button for 2 sec. The product indicator starts flashing blue.

Bluetooth pairing...

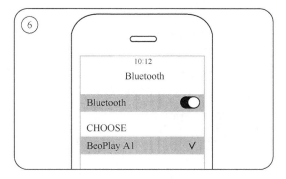

When the product indicator flashes blue, switch on Bluetooth on your device. Find the device list and select BeoPlay A1. The product indicator turns solid white, a sound prompt is heard and BeoPlay A1 is ready to play.

175

Connect

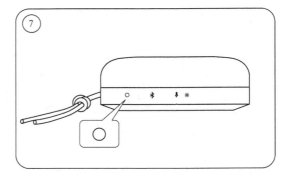

Use the Connect button to access your favourite feature with one press. Customize the button in BeoPlay app.

Speakerphone

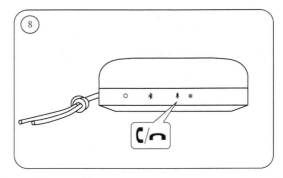

Short press to receive incoming calls. Long press to reject or terminate a voice call.

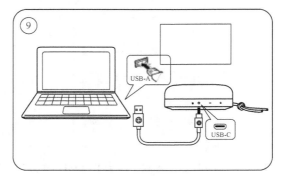

Connect your BeoPlay A1 to your computer with a USB cable, and then you can use it for conference calls.

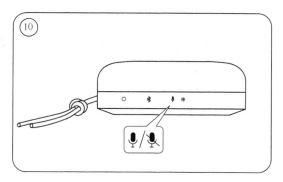

Short press to mute/unmute microphone during a call. The product indicator turns red/green. Double press to transfer audio to device or back to BeoPlay A1.

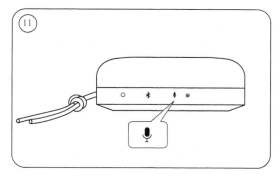

Long press to activate voice recognition.

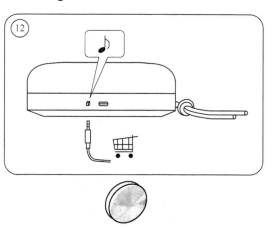

（节选自某公司产品说明书）

Chinese Translation of Texts（参考译文）

BeoPlay A1 用户指南

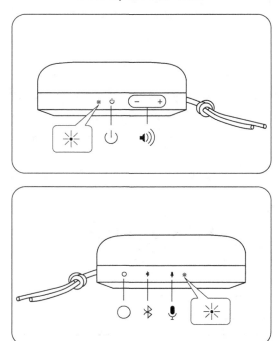

充电

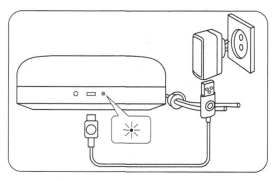

使用提供的 USB 线充电时，充电时间约为 3 小时。

电池指示器

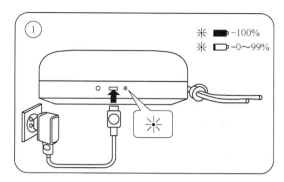

电池充电时,电池指示灯闪烁橙色。
电池充满后,电池指示灯变成绿色。

电池指示器

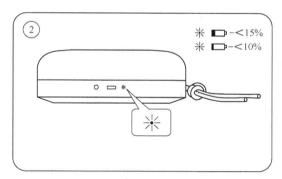

当电池剩余电量低于15%时,电池指示灯变成红色常亮。当电池剩余电量低于10%时,电池指示灯开始闪烁红色。

蓝牙配对……

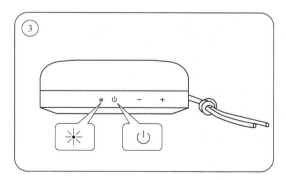

按下电源按钮,开启BeoPlay A1,产品指示灯变成白色常亮。此时,可以开始设置BeoPlay A1。

蓝牙配对……

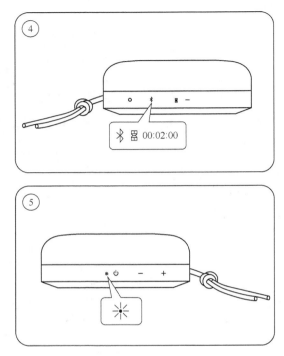

按住 Bluetooth 按钮 2 秒，产品指示灯将开始闪蓝光。

蓝牙配对……

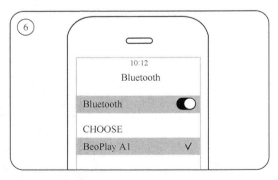

产品指示灯闪蓝光时，开启你设备的 Bluetooth 功能，搜寻设备列表并选择 BeoPlay A1。产品指示灯变成白色常亮，并听到声音提示，便可以开始播放 BeoPlay A1。

Unit Ⅴ Basic Knowledge of Professional English Ⅰ 专业英语基础知识 Ⅰ

连接

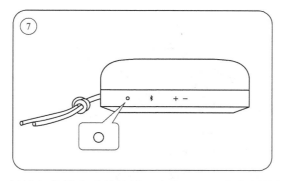

按下"连接"按钮一键访问你最喜欢的特性。在 BeoPlay 中定制按钮。

扬声器电话

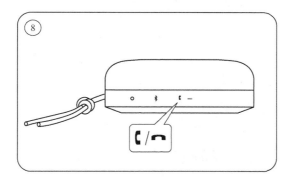

短按此按钮以接听来电。长按此按钮可拒绝或终止语音电话。

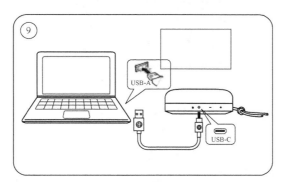

使用 USB 线将 BeoPlay A1 连接至计算机，然后你可以通过 BeoPlay A1 拨打会议电话。

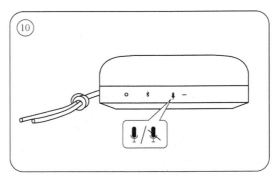

短按此按钮可在呼叫过程中关闭/打开麦克风。产品指示灯将变成红色/绿色。

连续两次按下此按钮可将音频传送至设备或传送回 BeoPlay A1。

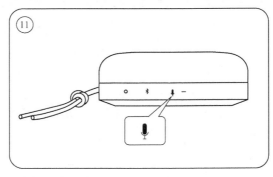

长按此按钮可激活语音识别功能。

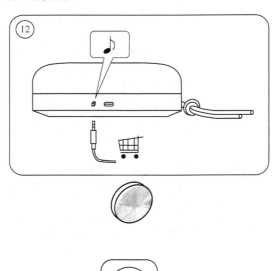

5.4.4　产品说明书资料 4

LFP power Li-ion Battery

1. Scope of Application

This specification describes the basic performance, technical requirement, testing method, warning and caution of the LFP power Li-ion rechargeable battery. The specification only applies to SHENZHEN JOOLEE BATTERY CO., LTD.

……

4. Performance Criteria

……

(3) Standard Charging Method

At 20±5℃, 0.3C5A constant current charge to 87.6V, then constant voltage 87.6V charge till charged current declines to 0.05C5A or less.

(4) Appearance

There shall be no such defect as scratch, flaw, crack, rust, leakage, which may adversely affect commercial value of battery.

……

7. Safe Characteristic

No.	Item	Test Methods and Condition	Criteria
(1)	Thermal shock	Put the battery in the oven. The temperature of the oven is to be raised at 5±2℃ per minute to a temperature of 130±2℃ and remains 30 minutes	No fire, no explosion
(2)	Over-discharge	The battery will be discharge with constant current 0.2C5A to cut-off voltage, then connect with external load of 120 ohm for 24 hours	No explosion, no fire, no smoke, no leakage
(3)	Short-circuit	After standard charging, connect batteries' anode and cathode by wire which impedance less than 100mΩ, keep 1 hours	No explosion, no fire, no smoke, no leakage
(4)	Over-charged	Charging cell with constant current 0.2C5A to voltage 43V, then with constant voltage 87.6V till current decline to 0. Stop test till batteries' temperature 10℃ lower than max temperature	No fire, no explosion

8. Storage and Shipment Requirement

Item		Criteria
Storage temperature	Short period less than 1 month	−10℃~45℃
	Long period less than 3 month	−10℃~35℃
	Long period more than 3 month	0℃~30℃
Relative Humidity		≤75% RH
Charged		About 40%~60% charged state

Battery must charge every three months. Long time store, please charge use 0.3C current keep 2 hours ensure the battery keep 40%~60% electricity.

9. Warning and Caution

(1) Do not put the battery into a fire or heat the battery; do not store the battery in high temperature environment.

(2) Do not connect the battery reversed in positive (+) and negative (−) terminals in the charger or equipment.

(3) Do not let the battery terminals (+ and −) contact a wire or any metal with carried or stored together, may cause short-circuit.

(4) Do not drive a nail in, hit with a hammer, or stamp on the battery.

(5) Do not disassemble or alter the batteries' outside structure.

(6) Do not submerge the battery in water, do not wet the battery when store the battery.

(7) Do not charge the battery continue over 24 hours.

Look out

(1) Do not use the battery direct sunlight, may cause battery overheating then fire or invalidation.

(2) Battery should be charged with proper charger, in compliance with correct operation contents.

(3) Do not discharge the battery continuously when it is charges.

（节选自某公司产品说明书）

Chinese Translation of Texts（参考译文）

磷酸铁锂动力电池说明书

1. 适用范围

本说明书规定了磷酸铁锂动力锂离子可充电电池的基本性能、技术要求、测试方法及注意事项。本说明书只适用于深圳市聚力电池有限公司所生产的磷酸铁锂动力锂离子电池。

……

4. 性能测试

……

（3）标准充电方法

在 20±5℃，以 0.3C5A 恒流充电至电压 87.6V，再以 87.6V 恒压充电至电流 0.05C5A 或更小。

（4）外观

电池外表面清洁，无电解液泄漏，无明显的划痕及机械损伤，无变形，无影响电池价值的其他外观缺陷。

……

Unit Ⅴ Basic Knowledge of Professional English Ⅰ 专业英语基础知识Ⅰ

7. 安全性能

序号	项目	测试方法和条件	标准
（1）	热冲击	将电池放进烘箱内，以 5±2℃每分钟的速度升高烘箱内温度至130±2℃后，恒温30分钟	不起火，不爆炸
（2）	过放	以 0.2C5A 恒流放电至截止电压，然后外接 300 Ω 负载放电 24 小时	不爆炸，不起火，不冒烟，无漏液
（3）	短路	将标准充电后的电池的正负极用一根小于 100mΩ 的导线连接，放置 1 小时	不爆炸，不起火，不冒烟，无漏液
（4）	过充	单体电芯用 0.2C5A 电流充电至电压 43V，然后恒压 87.6 V 让电流下降接近为 0，监视电池温度变化，当电池温度下降到比峰值低约 10℃时，停止实验	不起火，不爆炸

8. 存储及运输要求

项目		标准
存储温度	短期少于 1 个月	−10℃ ~ 45℃
	中期少于 3 个月	−10℃ ~ 35℃
	长期超过 3 个月	0℃~30℃
相对湿度		≤75%RH
荷电		约 40%~60%荷电状态

电池在贮存期间每三个月充电一次。长时间储存时，请客户对电池用 0.3C 的电流充电 2 小时，确保电池保持 40%～60%电量。

9. 警告和注意

（1）禁止将电池放入火中或者加热电池，不要将电池储存在高温环境中。
（2）禁止电池正负极反接，或直接将电池插入电源插座。
（3）禁止将电池的电极和导线或者其他金属物质接触及储存在一起，以免发生短路。
（4）禁止钉刺、敲击、抛掷、脚踩电池。
（5）禁止私自拆卸电池或取出电池外包装。
（6）禁止将电池放入水中，保存过程中应放置在阴凉干燥的环境中。
（7）禁止连续充电超过 24 小时。

*小心

（1）禁止在炙热的阳光下使用，可能会引起电池过热、起火或功能失效。
（2）电池需要采用合适的充电器，采用说明书建议的方式连接。
（3）电池在充电的过程中不允许持续放电。

5.4.5 产品说明书资料5

How to set up Apple AirPrnit

Apple AirPrint allows you to print e-mails, photographs and web-pages from iPad, iPhone and iPod touch using wireless LAN connection.

The Canon printer and the Apple product such as iPod, iPhone and iPod touch must be connected to the same IEEE802.11 wireless LAN platform to use Apple AirPrint.

Place the printer near the access point and use the printer function to set up wireless.

- Refer to the local Canon Web-site for how to print using Apple AirPrint.

1. Connection Using the WPS Function

(1) Press and hold the Maintenance button on the Operation Panel of the printer.

(2) Check to see if the LED flashes as illustrated.

(3) Press the Fit to Page button.

(4) Confirm that the LED flashes as illustrated, and then press and hold the WPS button on the access point for about five seconds.

- Refer to your access point manual for how to press the WPS button.

The Wi-Fi lamp (blue) of the printer flashes during searching and connecting the access point.

(5) When wireless LAN connection is completed, after the LED is lit up as illustrated for about three seconds, and then numeric "1" will be displayed.

Check to see if the Wi-Fi lamp (blue) of the printer is turned on at the same time.

2. Errors Appear on the LED

When an error is detected, "E" and the number flash on the LED of the printer.

(1) Press the Start button of the printer.

(2) After a while, repeat the operation following the instructions in this manual.

- If the error still occurs, turn off and on the access point, and then repeat the operation following the instruction in this manual.

（节选自某公司的产品说明书）

Chinese Translation of Texts（参考译文）

如何设置 Apple AirPrint

Apple AirPrint 允许你通过无线 LAN 连接，从 iPad、iPhone 和 iPod touch 打印电子邮件、照片及网页。

如需使用 Apple AirPrint，需要将 Canon 打印机及 Apple 产品（如 iPad、iPhone 和 iPod touch）连接至同一 IEEE802.11 无线 LAN 平台。

将打印机置于访问点附近，利用打印机功能，按如下所示即可设置无线 LAN。
- 有关如何利用 Apple AirPrint 进行打印的事宜，请参阅相应语言的 Canon 网站。

1. 使用 WPS 功能进行连接

在需参见图示的步骤中，请参见相应英文部分的图示。

（1）按住打印机操作面板上的维护按钮。

（2）检查 LED 是否如图中所示闪烁。

（3）按自动比例按钮。

（4）确认 LED 如图中所示闪烁，然后按住访问点上的 WPS 按钮大约 5 秒。

- 有关如何按 WPS 按钮的事宜，请参阅访问点手册。

在搜索及连接访问点的过程中，打印机的 Wi-Fi 指示灯（蓝色）将闪烁。

（5）完成无线 LAN 连接时，待 LED 如图中所示亮起约 3 秒后，将显示数字"1"。

检查打印机的 Wi-Fi 指示灯（蓝色）是否同时亮起。

2. LED 上显示错误

检测到错误时，打印机的 LED 上将闪烁"E"及数字。

（1）按打印机的启动按钮。

（2）稍等片刻，然后按照本手册中的说明重复相应的操作。

- 如果错误仍然发生，请关闭并重新打开访问点，然后遵照本手册中的说明重复相应的操作。

5.4.6 产品说明书资料 6

DELL

……

 CAUTION: General Safety Instructions

Use the following safety guidelines to help ensure your own personal safety and to help protect your equipment and working environment from potential damage.

Note: Additional user information for your computer, monitor, and individual components (such as storage drives, PC cards, and other peripherals) may be available under the "Manuals" section at support.dell.com.

Note: In this section, *equipment* refers to all portable devices (computers, port replicators, media bases, docking stations, and similar devices), desktop computers, and monitors. After reading this section, be sure to read the safety instructions pertaining to your specific equipment.

INPORTANT NOTICE FOR USE IN HEALTHCARE ENVIRONMENTS: Dell products are not medical devices and are not listed under UL or IEC 60601 (or equivalent). As a result, they must not be used within 6 feet of a patient or in a manner that directly or indirectly contacts a patient.

SAFETY: General Safety

When setting up the equipment for use:

- Place the equipment on a hard, level surface. Leave 10.2 cm (4 in) minimum of clearance on all vented sides of the computer to permit the airflow required for proper ventilation. Restricting airflow can damage the computer or cause a fire.
- Do not stack equipment or place equipment so close together that it is subject to recirculated or preheated air.

NOTE: Review the weight limits referenced in your computer documentation before placing a monitor or other devices on top of your computer.

- Ensure that nothing rests on your equipment's cables and that the cables are not located where they can be stepped on or tripped over.
- Ensure that all cables are connected to the appropriate connectors. Some connectors have a similar appearance and may be easily confused (for example, do not plug a telephone cable into the network connector).
- Do not place your equipment in a closed-in wall unit or on a bed, sofa, or rug.
- Keep your device away from radiators and heat sources.
- Keep your equipment away from extremely hot or cold temperatures to ensure that it is used within the specified operating range.
- Do not push any objects into the air vents or openings of your equipment. Doing so can cause fire or electric shock by shorting out interior components.
- Avoid placing loose papers underneath your device. Do not place your device in a closed-in wall unit, or on a soft, fabric surface such as a bed, sofa, carpet, or a rug.

（节选某公司产品说明书）

Chinese Translation of Texts（参考译文）

Dell 安全说明（节选）

……

 警告：一般安全说明

使用以下安全原则以帮助确保你个人的安全及保护设备和工作环境免受可能的损坏。

注：你的计算机、显示器和单个组件（如存储驱动器、PC卡和其他外围设备）的附件的用户信息可以在support.dell.com的"手册"部分得到。

注：在本节中，*设备*指所有便携式设备（计算机、端口复制器、Media Base、对接站及类似设备）、台式计算机和显示器。阅读本节后，请确保阅读适合您特定设备的安全说明。

在医疗环境中使用时的重要注意事项：Dell产品不属于医疗设备，未列在UL或IEC 60601标准（或同类标准）下。因此，不能在距病人6英尺的范围内使用这些产品，以直接或间接方式同病人接触都不行。

安全：一般安全

安装设备以进行使用时要注意以下几点。

- 将设备置于坚固水平面上。在计算机的所有通风孔侧保留至少 10.2 厘米（4 英寸）的空隙以保证良好的通风。通风不畅会损坏计算机或导致起火。
- 请勿堆叠设备或将设备靠得太近，因为这样会导致空气回流或空气预热。

注：将显示器或其他设备放到计算机上之前，请仔细审核计算机说明文件中涉及的重量限制。

- 确保没有任何物品压在设备的电缆上，且电缆所在的位置不会被踩到或导致绊跌。
- 请确保所有电缆连接到相应的连接器。有些连接器外观相近，容易混淆（例如，请勿将电话电缆插入到网络连接器中）。
- 请勿将设备放置到封闭的壁橱中及置于床、沙发或垫子上。
- 使设备远离暖气片和热源。
- 不要让设备在极热或极冷的温度下工作，确保其在指定的操作条件范围内工作。
- 请勿将任何物体放到通风孔中或设备开口处。这样做可能会导致内部组件短路，从而引起火灾或使人遭受触电。
- 避免在设备下方放置松散的纸张。请勿将设备放置到封闭的壁橱中，或置于柔软的织物表面上，如床、沙发、地毯或垫子上。

5.4.7 产品说明书资料 7

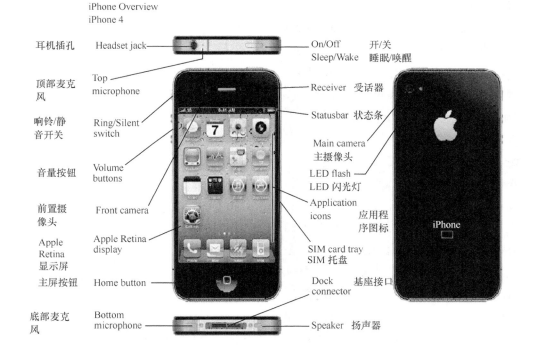

苹果手机说明书

5.4.8 产品说明书资料8

2SC1523 晶体管参数和特性（中英文版）

2SC1523 Transistor Datasheet Parameters and Characteristics

2SC1523　晶体管　参数和特性

Name: 2SC1523

名称：2SC1523

Material of transistor: Si

材料：硅晶体管

Polarity: NPN

极性：NPN 型

Maximum collector power dissipation (Pc): 1 W

最大集电极耗散功率（Pc）：1 W

Maximum collector-base voltage (Ucb): 40 V

最大击穿电压（Ucb）：40 V

Maximum collector-emitter voltage (Uce): 25 V

最大集电极-发射极电压（Uce）：25 V

Maximum emitter-base voltage (Ueb): 3 V

最大发射电压（Ueb）：3 V

Maximum collector current (Ic max): 300 mA

最大集电极电流（集成电路）：300 mA

Maximum junction temperature (Tj): 190℃

最大结温度（Ti）：190℃

Transition frequency (ft): 2 GHz

过渡频率（ft）：2 GHz

Collector capacitance (Cc), Pf: 2

集电极电容（Cc），Pf: 2

Forward current transfer ratio (hFE), min/max: 35/250

电流传输比（放大），最小/最大：35/250

Manufacturer of 2SC1523 transistor: HITACHI

生产 2SC1523：日立晶体管

Package of 2SC1523 transistor: TO128

2SC1523 晶体管封装：TO128

Application of 2SC1523 transistor: Ultra High Frequency, Medium Power

2SC1523 晶体管应用：超高频、中频电源

（节选某公司产品说明）

Unit VI

Basic Knowledge of Professional English II

专业英语基础知识 II

本单元介绍了专业英语的阅读（翻译）与写作，主要包括专业英语翻译的标准、专业英语翻译的过程、专业英语的基本特点、专业英语翻译的方法、专业英语的基本写作知识等。通过本单元的学习，学生可以了解专业英语的阅读（翻译）与写作的一些知识，提高专业英语阅读与写作的水平。

Professional English Reading (Translating) and Writing

专业英语阅读（翻译）与写作

6.1 专业英语翻译的定义

现代汉语词典对"翻译"一词的定义是，把一种语言文字的意义用另一种语言文字表达出来。专业英语翻译即是在准确、通顺的基础上，把一种专业语言信息转变成另一种专业语言信息的行为；同时也是把一种专业语言信息转变成另一种专业语言信息的过程，是一种再创作。

6.2 专业英语翻译的标准

毕生从事西方社会科学翻译的严复在《天演论》的序中说："译事三难：信、达、雅。"信，指译文要忠实准确地传达原文的内容。达，指译文要通顺流畅。雅，指译文要有文采，要注意修辞。

这三条标准对后世的翻译实践起到了巨大的指导作用，不过一百多年来，中国翻译界对它的评价褒贬不一，特别是对其中的"雅"字多有异议。

由于专业科技文体与其他文体在总体功能上有明显不同，专业英语注重表现技术问题的科学性、逻辑性和严密性，因此其翻译标准也有所侧重。一般说来，专业英语的翻译更注重"信"和"达"，即准确和通顺。

6.3 专业英语翻译的过程

对于翻译过程，翻译界有着不同的理解。我们认为，翻译过程可分为三个阶段：**理解、表达和校核**（校验）。

当我们进行翻译时，首先要阅读原文，理解作者的思想，然后根据自己的理解，用汉语表达出来。可见翻译包括理解和表达两个过程。翻译是复杂的脑力劳动，一般来说不能一蹴而就，通常为了保证翻译的质量，还应该仔细反复地校核和修改译文。

6.3.1 原文理解阶段

阅读理解是翻译过程的第一步，透彻地理解原文是确切表达的前提，阅读理解必须从整体着眼，联系上下文，结合专业背景知识，从专业角度贴近作者，在时间和空间上靠近作者，并考虑文化上的差异，力求与原作者达到心灵相通。

在阅读过程中弄清词汇及语法关系，尤其对长句、难句要认真，彻底辨明词义，弄清每个词语、词组、单句的确切含义。

6.3.2 汉语表达阶段

表达（即转换和重组）是翻译过程的第二步，是译者把自己所理解的内容正确、充分而又自然地传达给译文读者的过程。理解是表达的前提，表达是在理解的基础上进行的。

用汉语表达专业英语文献的内容时要注意用词简洁准确。一般来说，在表达阶段有两点必须注意。

1. **注意表达的规范性**

科技语言讲究论证的逻辑性、严谨性，要求语言规范，用词准确，造句恰当；不可有结构失调、指代不明、搭配不适当、累赘别扭等语病。

2. **注意表达的逻辑性**

逻辑性是专业英语的一个重要特点，不仅在理解原文时要注意事理分析和逻辑判断，在表达时也要注意这个问题，要用符合逻辑的语言传达原作的意图和思想，避免发生歧义。

6.3.3 检查校核阶段

检查校核是对理解和表达的进一步深化，是使译文符合标准的必不可少的一个阶段。在这个阶段，一方面要进一步核实原文的内容，另一方面要进一步推敲译文语言。

译者应具备的背景：至少通晓两种语言；较高的双语水平及相关文化知识；一定的专业技术知识，如电子信息、电子与通信工程等；还必须懂得翻译的理论、技巧和方法。

6.4 专业英语的基本特点

专业英语是一种重要的英语文体。作为独特的文体，它虽然有自身的特点，但并不意味着它是独立于普通英语之外的一种语言，其特点主要表现于频率和分布上，也就是说某些语言现象出现的次数比其他的语言现象多。

与非专业英语文体相比，专业英语具有词义多、长句多、被动句多、词性转换多、非谓语动词多、专业性强等特点，这些特点都是由专业文献的内容所决定的。

6.4.1 缩略词多

大量使用缩略词是专业英语的又一特点。缩略词趋向于任意构词，因而给翻译工作带来了一定的困难。

随着现代科学的高速发展，缩略语将日益增多。往往一个缩略语可代表几十个词义。在科技英语中缩略语的构成方式有以下三种。

第一种是首字母缩略词（Acronym/Initialism），由短语中每个单词的第一个字母构成，多用大写字母，可以拼读或用字母读。如可编程逻辑控制器（Programming Logic Control，PLC）（首字母组成的缩略词）、I/O（Input/Output）（首字母组成的缩略词）、计算机辅助设计（Computer Assistance Design，CAD）（首字母组成的缩略词）。

第二种是截短词（Clipped Word），缩略或截取自然词的一部分字母，可以是词首、词尾或词腰，从而形成缩略词。如 phone（telephone）、CPD（compound）、telex（telegram exchange）。

第三种是以辅音为核心组成的缩略词。这种缩写法主要用于缩写单词，可用大写字母，也可以用小写字母，一般用字母读，也可以拼读，如 Gds（Goods）商品。

除了一些常用的缩略词外，有时某篇论文的作者还会将仅在该文中使用的术语转换为缩略词。如交流电（Alternate Current，AC）、集成电路（Integrated Circuit，IC）、转/分（revolutions per minute，r.p.m）、场效应管（Field Effect Transistor，FET）、中央处理（Central Processor Unit，CPU）、图（Figure，Fig.）、调制解调器（MOdulator & DEModulation，MODEM）、等式（Equation，Eq）、万维网（world wide web，www）。

同时，英语中还时时都在产生各类缩略词。例如，从 radio detection and ranging 形成 radar（雷达），mechanical 和 electronics 组成 mechatronics（机电一体化）等。

6.4.2 词性转换多

英语单词中有不少是多性词，即既可用作名词，又可用作动词、形容词、介词或副词，字形无殊，功能各异，含义也各不相同，如不仔细观察，必致谬误，如 light。

名词：（启发）in (the) light of 由于，根据

　　　　（光）high light(s) 强光，精华

　　　　（灯）safety light 安全指示灯

形容词：（轻）light industry 轻工业

　　　　（明亮）light room 明亮的房间

　　　　（淡）light blue 淡蓝色

　　　　（薄）light coating 薄涂层

动词：（点燃）light up the lamp 点灯

副词：（轻快）travel light 轻装旅行

　　　（容易）light come, light go 来得容易去得快

193

诸如此类的词性转换在科技英语中屡见不鲜，几乎每个技术名词都可转换为同义的形容词。词性转换增加了英语的灵活性和表现力，读者必须从上下文判明用词在句中是何种词性，含义如何，才能正确无误地理解全句。

6.4.3 非谓语动词多

英语的每个简单句中，只能用一个谓语动词，如果读到几个动作，就必须选出主要动作当谓语，而将其余动作用非谓语动词形式，才能符合英语语法要求。

非谓语动词有三种：动名词、分词（包括现在分词和过去分词）和不定式。

例如：要成为一个名副其实的内行，需要学到老。

译文：To be a true professional requires lifelong learning.

这句中，有"成为""需要""学"三个表示动作的词。

可以看出，选择"需要"（require）作为谓语，其余两个动作中，"成为"用不定式形式 to be，而"学"用动名词形式 learning，这样才能符合英语语法要求。

使用非谓语动词可使文章简洁、醒目。

6.4.4 被动语态多

英语使用被动语态的语句大大多于汉语，科技英语更是如此，有三分之一以上的语句会用被动语态。

例如：No work can be done without energy.

译文：没有能量就不能做功。

例如：All business decisions must now be made in the light of the market.

译文：所有企业现在必须根据市场来做出决策。

科技文章的主要目的是讲述客观现象、介绍科技成果等，无须指出行为的主体，使用被动句比使用主动句有更少的主观色彩。因此在专业英语中，凡是在不需要或不可能指出行为主体的场合，或者在需要突出行为客体的场合，都使用被动语态。

例如：These parts of are usually connected by an electronic component referred to as a bus.

译文：这些部分通常是由被称为总线的电子组件连接的。

6.4.5 大量使用后置定语

后置定语在句中可以充当定语，对名词起修饰、描绘作用，还可以充当表语、宾语补足语等。形容词作定语修饰名词时，一般放在被修饰的名词之前，称作前置定语，但有时也可放在被修饰的名词之后，称作后置定语。

后置定语可以分为三大类：（1）定语从句；（2）短语，包括非谓语动词短语（不定式短语、现在分词短语、过去分词短语）、形容词短语、介词短语；（3）单个词。

专业文章要求行文简练，结构紧凑，往往使用分词短语代替定语从句或状语从句；使用分词独立结构代替状语从句或并列分句；使用不定式短语代替各种从句；使用介词+动名词短语代替定语从句或状语从句。这样可缩短句子，又比较醒目。

例如：One volt is defined as that magnitude of electromotive force required cause a current of one ampere to pass through a conductor having a resistance of one ohm.

译文：使 1 A 电流流过电阻为 1 Ω 的导体所需的电动势定义为 1 V。

（1）过去分词 required 充当后置定语，修饰 electromotive force。（2）having a resistance of one ohm 是现在分词短语，充当后置定语，修饰 conductor。

例如：…which produce large changes in the collector current flowing through the load.

译文：在负载中流动的集电极电流产生大的变化。

（1）which 引出 base current。（2）flowing through the load 是现在分词短语，充当后置定语，修饰 collector current。

6.4.6　常用定语从句

一个从句跟在一个名词或代词后进行修饰限定，从句在整个句子中做定语，这个从句就叫作定语从句。被修饰的词叫先行词。定语从句不同于单词作定语的情况，它通常只能放在被修饰的词（即先行词）之后。

定语从句由关系词（关系代词、关系副词）引导，关系代词、关系副词位于定语从句句首。

例如：The load converts these current changes into voltage changes which form the alternating output voltage u_o (u_o being much greater than u_i).

译文：负载将交变电流转变成交变电压即为输出交流电压 u_o（u_o 比 u_i 大）。

（1）to convert … into …：把……变成……，相当于 to change … into….（2）which form the alternating output voltage u_o 是定语从句，修饰 voltage changes。（3）alternating 是现在分词短语，修饰 output voltage。

例如：As a result a small alternating current is superimposed on the quiescent base current I_B which in effect becomes a varying DC.

译文：结果是一个小的交流电流叠加在静态基极电流 I_B 上并产生一个变化的直流电压。

（1）as a result："结果"。（2）which in effect becomes a varying DC 是定语从句，修饰 the quiescent base current I_B。

6.4.7　常用 it 作形式主语

例如：Often it is a small alternating voltage that has to be amplified.

译文：通常小的交流电压需要放大。

这是一个常见句型，it 是形式主语，that has to be amplified 引导真正意义上的主语从句。

例如：When it is desirable to express a magnitude of current smaller than the ampere,…

译文：当希望表述小于安培的电流大小时……

it 是形式主语，不定式 to express a magnitude of current smaller than the ampere 充当真正的主语。

6.4.8　复杂长句较多

科技文章要求叙述准确，推理严谨，因此一句话里包含三四个甚至五六个分句的并非少见。译成汉语时，必须按照汉语习惯破成适当数目的分句，才能条理清楚，避免洋腔洋调。这种复杂长句居科技英语难点之首，要学会运用语法分析方法来加以解剖，以便以短代长，化难为易。

专业英语用于表达科学理论、原理、规律、概述及各事物之间错综复杂的关系，而复杂的科学思维是无法使用简单句来表达的，所以语法结构复杂的长句较多地应用于专业英语，而这种严谨周密、层次分明、重点突出的语言手段也就成了专业英语文体的又一重要特征。

为了精确描述事物，以及表述一个复杂概念，并使之逻辑严密，结构紧凑，科技文章中往往会出现许多长句。有的长句多达七八十个词。

例如：For those of us who grew up at a time when the space age was not a part of everyday life, satellite-based communication is the culmination of a dream that stretches back to an era when the term satellite was only an idea conceived by a few inspired individuals.

译文：对于我们中那些并非生长在太空时代的人们来讲，卫星通信是人们长期以来一种梦想的顶点，这个梦想可以一直追溯到卫星这个词只是几个天才头脑中灵感的想象的那个年代。

由于专业英语著作具有很强的科学性和技术性，因此要求叙述准确、推理严密，而专业英语的表述对象是客观事物的发展过程、演变规律、影响因素、内在机理等，这些内容有时是相互关联、相互制约的，因此专业英语中复杂的长句十分常见，这是它与普通英语的一个显著区别。

6.5 专业英语翻译的方法和技巧

翻译是一种语言表达法，是译者根据原作者的思想，用另一语言将其文意表达出来。

6.5.1 专业英语翻译应遵循的原则

1. 忠实

译文应忠实于原文，准确地、完整地、科学地表达原文的内容，包括思想、精神与风格。译者不得任意对原文内容加以歪曲、增删、遗漏和篡改。

2. 通顺

译文语言必须通顺，符合规范，用词造句应符合译文语言的习惯，要用民族的、科学的、大众的语言，以求通顺易懂。不应有文理不通、逐词死译和生硬晦涩等现象。

专业英语翻译要求译者必须确切理解和掌握原著的内容和意思，丝毫不可以离开它而主观地发挥译者个人的想法和推测。在确切理解的基础上，译者必须很好地运用译文语言把原文通顺而流畅地表达出来。

译者是翻译的主体——译者在翻译活动中处于核心地位，起着中坚（而不仅仅是中间）作用。应遵循的原则：既尊重译者的主体性，又强调译者的客观性。

要比较圆满地解决这个问题，译者应具备以下能力：

（1）具有较扎实的专业知识；

（2）具有较好的英语基础；

（3）具有较好的汉语修养；

（4）还必须懂得翻译的理论、技巧和方法。

在专业翻译中，要达到融会贯通，必须了解相关的专业，熟练掌握同一事物的中英文表达方式。单纯靠对语言的把握也能传达双方的语言信息，但运用语言的灵活性，特别是选词的准确性会受到很大限制。要解决这个问题，翻译人员就要积极主动地熟悉专业领域的相关翻译知识。例如，要翻译"conductor"这个词，仅仅把字面意思翻译出来还远远不够，而且有时用词也不够准确。"conductor"在日常生活中的意思是"售票员和乐队指挥"，但在电学中却表示"导体"。了解了专业领域，在翻译过程中对语言的理解能力和翻译质量就会大大提高。

196

6.5.2 专业英语翻译的方法——直译与意译

直译指翻译过程中不仅要把原文的意思完整而正确地表达出来，而且要基本上保留原文的语言形式，如词语、句子结构、修辞手法等。

意译是指不拘泥于原文的表达形式等，在正确理解原文的基础上，重新遣词造句，把原文的意思用通顺的译文表达出来。

在翻译实践中完全直译或者完全意译的情况并不多见。通常的做法是直译和意译兼容并蓄，相互补充、相得益彰。

动笔前应将全篇文献通读一遍，对全文大意有个总的概念，然后开始逐字、句、段地翻译。如果一开始翻译就把注意力集中在一词一句的推敲分析上，就会只见树木不见森林，很可能出现错误。

动笔之前，对原文的语篇类型和文体特征进行认真的分析，是翻译理解阶段宏观把握原文的必然诉求。

1. 语言分析

语言分析主要包括对原文词汇意义（如一词多义、多词同义等）、各成分之间的语法关系（如主谓结构、修辞结构等）、修辞手段和惯用法（习语）等进行的分析。

这是翻译理解活动的起点。

2. 语篇类型及特征分析

语篇类型及特征分析主要包括对原文题材、体裁、文体、风格乃至语篇内在的衔接、连贯等的分析。

这对于我们确定翻译策略、选择翻译方法都是至关重要的。不同的语篇类型需要不同的翻译策略和方法与之相适应。

3. 对英语原文的分析和理解

实践证明：翻译之成败，理解是关键。要正确透彻地理解原文，首先必须根据原文的句子结构，弄清每句话的语法关系，尤其应当注重语法分析，可采用"分解归类、化繁为简、逐层推进"的方法，反复对原句进行剖析，突出句子的骨架，这对获得正确的理解十分重要。

另外对于涉及交叉学科的不太熟悉、难度较大的英语文献，可采用"多词扫描，逐层推进"的方法。

第一遍阅读：了解原文的大致内容、所属学科方向及内容的广度和深度。

第二遍阅读：分析篇章、段落结构及重点、难点所在。

第三遍阅读：主要分析疑难句子的语法结构，弄清句子骨架，标注疑难生词。

第四遍阅读：有选择、有目的地查阅字典词典，弄清上下文和句子结构理解的疑难生词，最后仔细地、有重点地理解原文。

6.5.3 英汉翻译中的词语翻译

一篇译作成功与否在很大程度上取决于词语的选择是否恰当、得体。选择了恰当、得体的词语就可以为译文的准确、通顺、流畅铺平道路。

翻译时遇到的第一个问题是词义的选择，专业英语词汇由专业技术词汇、半专业词汇、书面非专业词汇组成。单纯的专业技术词汇的翻译不成问题。而半专业词汇往往词义多变，一词多义，一词多性，并且英语词汇和汉语词汇大多不是一一对应的。专业英语文献的词汇中，有些词在不同专业领域有不同的含义。

词义的理解和选择实际上是对英语单词在特定句子中的含义判断和确定。词义的判断和确定在很大程度上取决于上下文和语言环境。因此，词义的选择或确定应从词的类别、词的搭配、上下文联系和专业知识等方面着手。

1. 根据词类选择词义

选择某个词的词义，首先根据句法结构判明这个词在原句中应属于哪种词类，对于确定词义是有益的。

2. 根据场合选择词义

词的多义性也反映在词的运用场合上，往往同一个词用在不同的学科或专业中就具有不同的意义。因此，在判断英语单词的词义时，可根据英语著作阐述内容所涉及的学科来确定词义。

例如：The fourth power of three is eighty one.

译文：三的四次方是八十一。（数学）

例如：Energy is the power to do work.

译文：能量是做功的能力。（物理）

3. 按照习惯搭配词义

英语和汉语都有一套词的搭配习惯，这主要表现在形容词和名词，以及动词和名词的搭配上。因此，翻译时有必要按照汉语的搭配习惯来处理英语句子中的某些搭配词，这样才能得到通顺的汉语译文。

4. 联系上下文法，活用词义

英语中有不少连词具有多重含义，因此同义连词可引导不同种类的从句。

例如：while 　当……时候；虽然；然而

　　　as 　　虽然；随着；由于；像……一样。

此时，连词的词义必须通过上下关系及整个句子来判断和选择。

5. 研究词义的引申

在词义的引申过程中，不仅要仔细研究上下文，有时还要考虑段落之间的关系。对于专业方面的内容，必须选用专业技术词语来表达。词义引申有 4 种不同的类型：（1）技术性引申；（2）修辞性引申；（3）具体化引申；（4）专业化引申。

6. 适当增词和减词

由于英语和汉语在遣词造句方面有差别，翻译时应注意在词量上有所增减。有时需要在译文中增加原文无其形而有其意的词。有时原文中的有些词在译文中不译出来，因为译文中虽无其词，但已有其意。

翻译时把英语和汉语的构词法和语义特性结合起来，采用音译、意译、形译、象译或音意结合的方法，可达到准确、简洁、通俗易懂。

缩略词或缩写词经常大量出现在科技文章中，掌握一些常见的缩略词对阅读外文资料是十分必要的。

6.5.4 英汉翻译中句子的翻译

根据专业英语语言的特点，在阅读、理解和翻译过程中要注意使用各种不同的技巧和方法，达到准确、生动、形象的目的。翻译技巧就是在翻译过程中用词造句的处理方法，如词义的选择、直译、意译、词类转换和专业术语的翻译方法等。

1. 被动语态的翻译

据英国利兹大学约翰·斯维尔斯（John Swales）的统计，专业英语中的谓语至少有三分之一是被动语态。

例如：Attention must be paid to the working temperature of the machine.

译文：应当注意机器的工作温度。

不一定要千篇一律地将英语的被动结构都译成汉语的被动结构，而应根据实际情况，"因地制宜"地进行翻译，使译文尽可能地符合汉语的表达习惯。常见的翻译方法如下。

（1）译成汉语的被动句。不改变原文语序，通过"被""由""为""受""用""给""叫""靠""让"等介词引出动作执行者。

（2）译成汉语的主动句。

（3）译成汉语因果句。由 by 或 in 引导的状语往往可以转换为汉语的主语。

2. 复杂长句的翻译

在专业英语中，为了表述一个复杂概念，使之逻辑严密，结构紧凑，常采用包含若干从句的复杂句或包含许多附加成分的简单句。因此，与普通英语相比，专业英语所使用的长句较多。至于难句，不是语法上的概念，只是翻译起来较为复杂的句子，是指含有几个错综复杂关系的复合句。

例如：Factories will not buy machines unless they believe that the machine will produce goods that they are able to sell to consumers at a price that will cover all cost.

这是由一个主句和四个从句组成的复杂长句，只有进行必要的语法分析，才能正确理解和翻译。现试译如下。

译文：除非相信那些机器造出的产品卖给消费者的价格足够支付所有成本，否则厂家是不会买那些机器的。

节译：要不是相信那些机器造出的产品售价够本，厂家是不会买的。

后一句只用了 25 个字，比前句 40 个字节约用字 37.5%，而对原句的基本内容无损。可见，只要吃透原文的结构和内涵，翻译时再在汉语上反复推敲提炼，复杂的英语长句也是容易驾驭的。

在翻译长句时，只要充分理解原文内容，特别是句法分析，并掌握一定的翻译技巧，就可以化难为易了。

3. 英语长句的分析

一般来说，造成长句的原因有 3 个方面：（1）修饰语过多；（2）并列成分多；（3）语言结构层次多。

在分析长句时可以采用以下方法。

（1）找出全句的主语、谓语和宾语，从整体上把握句子的结构。

（2）找出句中所有的谓语结构、非谓语动词、介词短语和从句的引导词。

（3）分析从句和短语的功能。例如，是否为主语从句、宾语从句、表语从句等，若是状语，它是表示时间、原因、结果，还是表示条件，等等。

（4）分析词、短语和从句之间的相互关系。例如，定语从句所修饰的先行词是哪一个，等等。

（5）注意插入语等其他成分。

（6）注意分析句子中是否有固定词组或固定搭配。

例如：For a family of four, for example, it is more convenient as well as cheaper to sit comfortably at home, with almost unlimited entertainment available, than to go out in search of amusement elsewhere.

（1）该句的骨干结构为 it is more … to do sth. than to do sth. else，这是一个比较结构，而且是在两个不定式之间进行比较。（2）该句中共有 3 个谓语结构，它们之间的关系为，it is more convenient as well as cheaper to … 为主体结构，但 it 是形式主语，真正的主语为第二个谓语结构 to sit comfortably at home，并与第三个谓语结构 to go out in search of amusement elsewhere 作比较。（3）句首的 for a family of four 作状语，表示条件。另外，还有两个介词短语作插入语：for example，with almost unlimited entertainment available，其中，第二个介词短语作伴随状语，修饰 to sit comfortably at home。

译文：譬如，对于一个四口之家来说，舒舒服服地在家中看电视，就能看到几乎数不清的娱乐节目，这比到外面别的地方去消遣便宜和方便。

4. 理解全句

句子冗长、结构复杂是专业英语的特点之一。

翻译长句时，首先要能理解全句，其次要弄清两种语言的不同表达方式。

例如：However great the joy with which he welcomed a new discover in some theoretical science whose practical application perhaps it was as yet quite impossible to envisage, he experienced quite another kind of joy when the discovery involved immediate revolutionary changes in industry, and in historical development in general.

本句中，however 用作副词，起连词作用（注意，however 后无","或并列连词，不具有插入语性质），它引导让步状语从句。此句既有倒装，又有省略。倒装是指"great"受"however"影响放至前面；省略是指省略了 was（这个从句的正常语序应为 The joy was however great.，但英语习惯上并不这样说），以加强语势，并使主语 joy 与跟在它后面的定语从句连接得更紧密。这个定语从句（由 which 引导）中又包含了一个由 whose 引导的定语从句。对这两个从句都要先找出先行词，然后以先行词替换定语从句中的关系代词，这样理解意义就比较容易。第一个","后是主、谓、宾形式的主句，由 when 引导状语从句，修饰主句中的谓语动词 experienced，其中，用 and 引入的 in…短语，重复 in，使 in historical development 的地位更加突出。

译文：虽然任何理论科学中的每一个新的甚至尚无从预见其应用的发现，都能使他感到特别喜悦，但当有了立即会对工业、对整个历史发展起革命性影响的发现时，他所感到的喜悦就更加异乎寻常了。

专业英语中长句较多，在翻译时多采用拆译法，即在对全局语法关系、逻辑层次和时间先后进行分析的基础上，将长句中的并列成分、短语、从句及附加成分——拆译出来，按照汉语的表达习惯，对原文的句子结构和表达层次进行适当的调整和重新组合。

5. 确定主谓，拆开句子

在理解长句时，首先应确定主谓组数，将句子化"整"为"零"。即通过拆开句子，使其简化成简单句。而简单句则可通过中心词来理解其全句要点。

例如：When a body which has a certain temperature becomes hotter, we say that heat has been added to it.

译文：当具有一定温度的物体变得更热时，我们说有热量加到该物体上去了。

本句有4个谓语（has，become，say，has been added）和4个主语（body，which，we，heat），即共4个简单句。全句中有3个连词和关联词（when，which，that）。无连词和关联词的句子，显然是主句，其余为从句。when引导时间从句，which引导定语从句，that引导宾语从句。经过这样的剖析，这个长句就不难翻译了。

6. 通过连词，弄清拆开句子各部分之间的关系

连词分为并列和从属两种，前者连接并列句中的分句，后者连接复合句中的主句和从句。

6.5.5 专业英语翻译的注意事项

（1）多留意常用词语，小心掉进常用意义的陷阱中。

（2）勤查词典，注意一词多义。在英语中，一个词语常常是集多种意义于一身的，而在具体的上下文中却只有一个意义，这个意义是依赖其所在的上下文或者该词语同其他词语的搭配或组合关系而衍生出来的。我们在翻译中就要充分注意这种现象，勤查专业词典，将英语中所表达的真实意义传达出来。

（3）注意词语意义的感情色彩。在译文中选择词语还要考虑词语意义的感情色彩。这时的选词并不一定有对错之分，但有精确与否、得体与否的区别及优劣的区别。

（4）仔细区分词语的使用语体。在专业英语中所使用的语言一般来说都是比较客观的，专业文体崇尚严谨周密，概念准确，逻辑性强，行文简练，重点突出，句式严整，而且所使用的往往是客观的、不带非常强烈的个人倾向的语言。

（5）注意表达方式的调整，使译文符合逻辑和实际情况。

（6）注意表达的简洁性，使句子既简洁又达意。

6.6 专业英语写作知识

专业英语文章一般由导言、正文和结论（或结束语）3部分组成。当然语篇结构可以因陈述的内容、时间、场合、对象的不同而有所差异。学术论文的写作则另有要求。

6.6.1 专业英语学术论文写作

所谓学术，是指较为专门的、有系统性的学问，学术论文是学术成果的载体，它的内容是作者在某一科学领域中对某一课题进行潜心研究而获得的结果，具有系统性和专门性，而不是点滴所得。

学术论文应具有一定的理论价值，要提示事物的本质，反映客观规律。在写作中，作者需要用大量的可靠材料，运用科学的方法对本质的东西加以剖析，对规律性的东西进行探讨。学术论文的每篇论文只叙述一个观点，要准确描述所进行的研究的内容。

学术论文的作者要能用科学的思想方法进行研究，并得出科学的结论。科学性是学术论文的灵魂，没有科学性的"学术论文"是没有生命力的。

6.6.2 一般学术论文包括的几个部分

1. 标题、副标题（Title，Subtitle）

标题高度概括；准确得体；简短精练；醒目。

2. 姓名和单位（Author and Department）

3. 摘要（Abstract）

摘要一般应写出 4 点：（1）论文涉及的从事这一研究的目的和重要性；（2）研究的主要内容，指明完成了哪些工作及研究的方法；（3）获得的基本结论和研究成果，突出论文的新见解；（4）结论或结果的意义。

4. 关键词（Key words）

一篇论文可选取 3~8 个词作为关键词。关键词的一般选择方法是，由作者在完成论文写作后，纵观全文，选出能表示论文主要内容的信息或词汇，这些词语可以从论文标题中或从论文内容中去寻找和选择。

5. 目录（Catalog）

目录是全篇论文的提纲，含各章节的小标题，可帮助读者了解全文结构。目录取正文中的一、二级标题即可，不要写出三、四级标题。列出标题之后，要带上页码。

6. 引言（Introduction）

引言又称前言，属于整篇论文的引论部分。写作内容包括研究的理由、目的、背景、前人的工作和知识空白、理论依据和实验基础、预期的结果及其在相关领域里的地位、作用和意义。

引言的文字不可冗长，内容选择不必过于分散、琐碎，措辞要精练，要吸引读者读下去。引言的篇幅大小，并无硬性的统一规定，需视整篇论文篇幅的大小及论文内容的需要来确定，长的可达 700~800 字或 1000 字左右，短的可不到 100 字。

7. 论文主体（Subject of paper）

学术论文必须是有自己的理论系统的，不能只是材料的罗列，应对大量的事实、材料进行分析、研究，使感性认识上升到理性认识。它必须切实地从客观实际出发，在论据上，应尽可能多地占有资料，以最充分的、确凿有力的论据作为立论的依据。在论证时，必须经过周密的思考，进行严谨的论证。

要用通俗易懂的语言表述科学道理，不仅要做到文从字顺，而且要准确、鲜明、和谐、力求生动。

论文的组织方法是作者自己确定的，应按逻辑顺序组织材料。要强调主要观点，并加以适当的说明，不重要的观点则用适当的附属的方法表示（注意突出重点）。不要把论文写成泛泛而谈。

8. 结论（Conclusion）

结论要总结并评价研究工作的结果，指出这些工作的意义和优点及工作中的局限性和存在的问题，并描述未来的前景和应用。

正文要有页码，页码居中。如果有注释，最好用"尾注"的形式（但正规的论文是用"脚注"）；参考文献不要再作为注释。

9. 致谢（Acknowledgement）

致谢是指对论文有帮助的人表示的感谢。

10. 参考文献（References）

列出参考的书目及参考文献。参考文献格式为，序号、作者、书名（论文名）、出版社（期刊名）、出版时间（期刊时间）。

11. 附录（Appendix）

附录要列出文中所引用的图表或表格，以及所需要列出的资料。

6.7 专业英语学术论文常用句式举例

1. 写摘要常用的句式，介绍论文内容或作者的观点

例如：The authors present a…principle that predict…

译文：作者提出了一种……原理，该原理能预测……

例如：The paper presents the data in terms of…

译文：本文用……给出了……的数据。

例如：The authors describe a configuration of…

译文：作者描述了一种……构造。

例如：This paper describes the principles and techniques of…

译文：本文描述了……的原理和技术。

例如：Performance goals and various approaches to … are briefly described.

译文：简述了性能目标和达到……的各种途径。

例如：This paper addresses the problem of…

译文：本文论述……的问题。

2. 介绍科技论文研讨的课题

例如：The performance characteristics of…are studied theoretically and experimentally.

译文：对……的性能特征进行了理论和实验研究。

例如：The paper examines…and consider…

译文：本文研究了……，并考虑了……

例如：The authors consider two specific subjects which…

译文：作者考虑了两个……的专门课题。

例如：A principle of constructing…is considered in this paper…

译文：本文考虑了建立……的原则。

例如：This article discusses the reasons for… and offers an insight into…

译文：本文讨论了……的原因并提出了对……的观点。

3. 介绍论文涉及的范围

例如：Some of the specific topics discussed are:

译文：论述的专门课题包括……

例如：The article contains some practical recommendations on…

译文：本文包含了一些有关……的实际建议。

例如：The scope of the research covers…

译文：本研究的范围涉及……

例如：Subjects covered include…

译文：涉及的课题包括……

4. 综述与概括对某一领域的研究课题

例如：This paper reviews…, summarizes the theory from…viewpoint, discusses…, and presents…

译文：本文综述了……，从……观点概述了这一理论，并讨论了……，还提出了……

5. 介绍论文重点

例如：This report concentrates on…

译文：本报告的重点是……

例如：Attention is concentrated on…

译文：重点是……

例如：Particular attention is/was paid to…

译文：特别重视……

例如：There has been a focus…and attention is being paid to…

译文：……一直是重点，而且现在都重视……

例如：The greatest emphasis has been on…

译文：极其重视……

例如：The primary emphasis in this paper is on…

译文：本文的重点是……

6. 介绍论文的目的

例如：One of the purposes of this study is…

译文：本研究的一个目的是……

例如：The aim of this study is to carry out analysis for…

译文：本研究的目的是对……进行分析。

例如：The primary objective of the study was to determine…

译文：本研究的主要目标是确定……

7. 介绍论证与依据

常用词语有 be based on，base on，take a reference 等。

例如：The method is based on…

译文：这种方法的基础是……

例如：Constitutive equation of…is formulated based on…

译文：……的构成式是以……公式为基础的。

8. 介绍推荐与建议

常用的词语有 propose、suggest、recommend 等。

例如：The author proposes an approach to…

译文：作者建议创立一种方法来……

例如：Suggestions were made for further study of…

译文：提出了进一步研究……的建议。

例如：It is suggested that…

译文：有人建议……

9. 介绍结论的语句

常用词语有 conclude、arrive at 等。

例如：The paper concludes that…

译文：本文的结论是……

例如：It was concluded that…

译文：得出的结论是……

10. 正文常用的一些句型

（1）阐明论文目的（宗旨）时常用的句型

① 本文的主要目的是……

a. The purpose of this paper is…

b. The primary goal of this research is…

c. The intention of this paper is to survey…

d. The overall objective of this study is…

e. In this paper, …was provided（given, illustrated）…

② 该项工作是关于……

a. The present work is concerned with the processes underlying…

b. The above work deals with the mechanism involved in…

c. Our work at present is devoted to the changes occurring in…

d. The work we have done bears on the effects produced by…

e. The work we are doing is closely related to the deliberations described in…

③ 该项工作的主要目的是……

a. The chief aim of the present work is to investigate the feature of…

b. The main purpose of our recent work was to examine the properties of…

c. The major object of their further work will be research into the nature of…

d. Our work in this direction has been intended to provide some information about…

e. The work which is being done now is designed to find some solution to the problem of…

（2）指明研究范围时常用的句型

① 该问题是关于……的研究

a. This problem is concerned（chiefly）with the study of…

b. The problem they have advanced bears on the elucidation of…

c. The problem to be discussed in this paper is related closely with…

d. The problem described previously was directed to the example…

② 这是研究……的问题

a. This is a problem concerned with the nature of…

b. This is a problem relating to the influence of…

c. This is a problem which deals with the role of…

d. This is a problem which bears on the effect of…

e. This is a problem in connection with…

③ 该问题在……范围之内（之外）

a. The problem in/within the scope of…

b. Our problem lies beyond the range of…

c. The problem you have just outlined seems to be outside of the province of…

（3）阐述理论时的常用句型

① 该理论是……提出来的

a. The theory of…was created 50 years ago.

b. The alternative theory was constructed in 80's.

c. The theory of…was developed in the 1999's.

d. The theory of…was elaborated in the early 70's.

e. The theory of…was formulated as early as 2000.

② 该理论的内容（特点）是……

a. Underlying this theory is the idea that…

b. The underlying concept of this theory is…

c. The underlying principle of the theory is as follows…

d. This is an alternative theory that…

e. This is a similar theory to the effect that…

③ 根据该理论，可得出……

a. According to B's theory both processes occur simultaneously.

b. As follows from the theory by N, such situation lead to…

c. As can be seen from the theory advanced by…such interactions are the basis of…

d. In the light of the theory, we have developed a variational method to handle such problems.

e. Based on the theory, the author has obtained sufficient information from recent experiments with…

（4）介绍方法时常用的句型

① 该方法首先由……发明（利用）。

a. The method of…was first developed by N.

b. The method of…was applied by B.

c. The method of…was first used by A in the early XXth century.

d. The method of…was first brought into used by C.

e. The method of…was came into use as long as 2000.

② 该方法与其他方法有……差别。

a. The method of…does not differ at all the routine one.

b. The method we use differs greatly from the one we used earlier.

c. The method they developed is somewhat different from the conventional one.

d. The new technique is quite different from the old one.

e. The newly-elaborated technique has something（nothing，little）in common with the one previously used.

f. The method of deriving a program described here differs from previous practice in that…

③ 应用该方法可以进行……工作。

a. The method allows us to demonstrate…

b. This technique enables us to observe…

c. This method is capable of providing…

d. By the method we are able to investigate the insight of…

（5）展示结果时的常用句型

① 该项研究说明（指出、揭示）了……

a. These pioneer studies that the authors attempted have indicated marked variation in…

b. We carried out several studies which have demonstrated that the substance acts on…

c. The research we have done suggests an increase in…

d. The investigation carried out by…has revealed on resistance of…

② 根据该项实验结果可得出……

a. More recent experiments in this area lead us to conclude that the phenomenon is related to…

b. Most recent experiments to the same effect have led the authors to believe that the mechanism out of action.

c. Further experiments in this area have enable these workers to suggest that…

d. As a result of our experiments, we concluded that…

e. From our experiments, the authors come to realize that…

③ 该项工作提供了（没有得出）……结果。

a. This fruitful work gives explanation of…

b. Their fundamental work provides some knowledge of…

c. The author's pioneer work has contributed our present understanding of…

d. This joint work of many years has led to further progress in…

e. Our integrated work of the past few years adds to our current concept of…

（6）表示感谢的句型

例如：I would like to thank…

译文：我要感谢……

例如：The author wishes to express his sincere thanks to…

译文：作者愿对……表示由衷的感谢。

例如：The author wishes to make the acknowledgement of…

译文：作者愿感谢……

例如：Grateful acknowledgement is made to…

译文：对……深表感谢。

例如：The author is in acknowledgement of…

译文：作者要感谢……

例如：I acknowledge…

译文：我要感谢……

例如：Special gratitude is owed to…

译文：我要特别感谢……

例如：I am indebtedness to…

译文：我对……表示感谢。

6.8 专业英语中数字的表示方法和读法

6.8.1 整数

- 35,896,732,546,285

读作：thirty-five trillion eight hundred and ninety-six billion seven hundred and thirty-two million five hundred and forty-six thousand two hundred and eighty-five

注意在上述这样的数词中，hundred，thousand，million，billion，trillion 等词一般是单数形式。例如，two hundred，three million，five trillion，etc.

- 第 403 号房间

  Room 403

  读作：room four oh three

- 电话号码：0371-7426329

  Telephone Number: 0371-7426329

  读作：zero three seven one seven four two six three two nine

- 97 次列车

  Train Number 97

  读作：train ninety-seven 或 the ninety-seventh train

- 美国西北航空公司 3644 航班

  Flight NW 3644

  读作：flight NW three six four four

- 公元前 753 年

  （in）the 753 BC

  读作：（in）the year seven hundred and fifty-three BC

- 20 世纪 70 年代

  1970's 或 1970s

  读作：（in）nineteen seventies

- 1945 年 8 月

  in August, 1945

  读作：in August nineteen forty-five

- 1997 年 7 月 1 日

  on July 1(st), 1997 或 on 1(st) July, 1997

  读作：on July the first ninety-seven 或 on the first of July nineteen ninety-seven

- 上午九点十分

  09:10 a.m.

  读作：（oh）nine ten a.m.

6.8.2 小数和分数

- 329,671,239.953579

  读作：three hundred and twenty-nine million six hundred and seventy-one thousand two hundred and thirty-nine point nine five three five seven nine

- 2,050.0357

  读作：two thousand and fifty point zero three five seven

- .01

  读作：one hundredth（point zero one）

- .00001

  读作：one millionth（point zero zero zero zero one 或 point four zeros one）

- 9.621

  读作：nine point six two one

- $\dfrac{3}{7} = 0.428571428571$

  读作：Three divided by seven equals zero point four two eight five seven one four two eight five seven one.

- $5\dfrac{5}{8}$

  读作：five and five-eighths（five and five over eight 或 five and five divided by eight）

- $0.003\% = \dfrac{0.003}{100} = 0.00003$

  读作：Zero point zero zero three percent equals zero point zero zero three over one hundred equals zero point four zeros three.

- $0.25 = 0.25 \times 1000‰ = 250‰$

  读作：Zero point two five equals zero point two five times thousand per mille equals two hundred and fifty per mille.

6.9　一般数学符号的读法

1. 加法

2 + 3　　　　读作：two plus three

2. 减法

10 − b　　　　读作：ten minus b

3. 正负号

X = ± 5　　　　读作：X equals minus or plus five

4. 乘法

x · y　　　　读作：x times y 或 x multiplied by y

5. 除法

9 ÷ 4　　　　读作：nine divided by four

6. 比率

1 : 2　　　　读作：the ratio of one to two

a : b　　　　读作：a is to b

7. 分数

$\dfrac{1}{2}$　　　　读作：one-half

$\dfrac{2}{3}$　　　　读作：two-thirds

（1）百分数

0.3%　　　　读作：zero point three percent

（2）千分数

3.8‰　　　　读作：three point eight per mille

8. 平方

b^2　　　　读作：b squared

9. 平方根

$\sqrt{x}$ 读作：the square root of x

10. 乘方

记作 x^n，表示 x 的 n 次方。

读作：the nth power of x, or x to the nth power.

a^n 读作：a to the nth power

11. 开方

$\sqrt[n]{a} = b$

读作：The principal n^{th} root of a is b.

6.10 专业课程名称中英文对照

电路基础　　Fundamentals of Electric Circuit

模拟电子技术　　Analog Electronics Technology

数字电子技术　　Digital Electronics

模拟集成电路设计　　Analogue IC Design

智能控制　　Intelligent Control

可编程控制器系统　　Programmable Logical Controller System

电子电路设计　　Electronic Circuits Design

印制电路板设计　　Printed Circuit Board Design

电子测量技术　　Electronic Measurement Technology

高频电子技术　　High Frequency Electronic Technology

电子产品装配与调试　　Electronic Products Assembly and Debugging

单片机应用　　Application of the Single Chip Microcomputer

微机原理　　Microcomputer Principle and Interface Technology

计算机仿真　　Computer Simulation

无线传感器网络　　Wireless Sensor Networks

数字通信　　Digital Telecommunications

光纤通信　　Optical Fiber Communications

自动控制理论　　Automatic Control Theory

数字信号处理　　Digital Signal Processing

电子通信技术实验　　Experiments in Electronic Communication Techniques

通信与计算机网络　　Communications and Computer Networks

电磁场　　Electromagnetic Field

无人机系统　　Unmanned Aircraft Systems

短距离数据传输技术　　Short Distance Data Transmission Technology

射频识别技术与应用　　RFID Technology and Application

线性系统　　Linear System

楼宇自动化　Building Automation
电力电子技术　Power Electronics Technology
电力系统分析与设计　Power System Analysis and Design
专业英语　Professional English
高等数学　Advanced Mathematics
拓扑学　Topology
微积分　Calculus
传感器　Sensors

APPENDIX I
Vocabulary

词汇表

A

abridged [ə'brɪdʒd] adj.	削减的，删节的
abstract ['æbstrækt] n.	取出，吸收
access ['ækses] vt.	使用，接近，获取
accommodate [ə'kɔmədeit] v.	容纳
achieve [ə'tʃiːv] vt.	完成，达到
acquisition [ˌækwi'ziʃn] n.	获得，取得，获得物，【无线】探测
adaptive modulation	自适应调制
aerial ['eəriəl] n.	天线
aerodynamics [ˌɛərəudai'næmiks] n.	空气动力学，气体力学
aftermath ['ɑːftəmæθ] n.	结果，后果
aggregate ['ægrigət] vt.	使聚集，使积聚，总计达
alarm [əlaːm] vt. & n.	警告；报警器
alternating ['ɔːltəˌneitiŋ] adj.	交互的
altitude ['æltitjuːd] n.	（尤指海拔）高度
ammeter ['æmiːtə] n.	安培表
amorphous [ə'mɔːfəs] adj.	无定形的，无组织的，【物】非晶形的
ampere ['æmpɛə] n.	安培
amplification [ˌæmplifi'keiʃən] n.	放大

amplifier ['æmplifaiə] n.	放大器
amplitude ['æmplitju:d] n.	振幅
anode ['ænəud] n.	阳极，正极
antenna [æn'tenə] n.	天线
apparatus [ˌæpə'reitəs] n.	器械，设备，仪器
appearance [ə'piərəns] n.	出现，外观
approach [ə'prəutʃ] n. & vt.	接近
approaching [ə'prəutʃiŋ] vt.	接近，靠近
architect ['ɑ:kitekt] n.	建筑师
arrangement [ə'reindʒmənt] n.	装置
array [ə'rei] n.	队列，阵列
assistant [ə'sistənt] n.	助手，助理
astronomical [ˌæstrə'nɔmikəl] adj.	天文的，天文学的
attenuate [ə'tenjueit] v.	削弱
attenuator [ə'tenjueitə] n.	衰减器
automatic [ɔ:tə'mætik] adj. & n.	自动的，无意识的，必然的；自动机械，自动手枪
average ['ævəridʒ] adj.	平均的

B

bandpass ['bændpɑ:s] n.	带通
bandwidth ['bændwidθ] n.	带宽
barcode ['bɑ:kəud] n.	条形码，条码技术
base [beis] n.	基极
bias ['baiəs] v. & n.	偏置
biased ['baiəst] adj.	结果偏倚的，有偏的
bi-directional [baidi'rekʃən(ə)l] adj.	双向的，双向作用的
bill [bil] n.	账单，目录
binary ['bainəri] adj.	二进位的，二元的
block [blɔk] n.	街区
broadcasting ['brɔ:dˌkɑ:stiŋ] n.	广播
built-in ['bilt'in] adj.	嵌入的，内置的
bus [bʌs] n.	【计】（电脑的）总线

C

cadmium telluride	碲化镉
calamity [kə'læməti] n.	灾难，不幸事件

calendar ['kælində] n.	日历
capacitor [kə'pæsitə] n.	电容器
capitalize ['kæpitəlaiz] vi.	利用
categorize ['kætigəraiz] v.	加以类别，分类
category ['kætəg(ə)ri] n.	类别，分类
cathode ['kæθəud] n.	阴极
ceramic [si'ræmik] adj. & n.	陶器的；陶瓷
characteristic [ˌkærəktə'ristik] n.	特性，特征
charge [tʃɑ:dʒ] n.	电荷
chip [tʃip] n.	芯片
circuit ['sə:kit] n.	电路
cockpit ['kɔkpit] n.	（飞机驾驶员的）驾驶舱，座舱
collector [kə'lektə] n.	集电极
collision [kə'liʒn] n.	碰撞，冲突
commercially [kə'mə:ʃəli] adv.	商业上地
commonsense ['kɔmən'sens] adj.	具有常识的
compatible [kəm'pætəbl] adj.	兼容的
complementary ['ʌltimət] adj.	补充的，补足的
component [kəm'pəunənt] n. & adj.	成分，元件；组成的，构成的
comprise [kəm'praiz] vt.	包含，包括，由……组成
comprise [kəm'praiz] v.	包含，由……组成
compromise ['kɔmprəmaiz] n.	妥协，折中
conductor [kən'dʌktə] n.	导体，导线
configuration [kənˌfigju'reiʃən] n.	结构
confusing [kən'fju:ziŋ] adj.	使人困惑的，令人费解的
console [kən'səul] n.	控制台，操纵台
constraint [kən'streint] n.	约束，限制
contradiction [kɔntrə'dikʃn] n.	矛盾，否认，反驳
conversion [kən'və:ʃn] n.	变换，转化
convert [kən'və:t] vt.	使转变，转换
converter [kən'və:tə] n.	变换器，换流器，变压器，变频器
convey [kən'vei] vt.	搬运，传达，转让
copper indium gallium sulfide	铜铟镓硫化物
corresponding [ˌkɔ:ri'spɔndiŋ] adj.	符合的，一致的，相同的
coulomb ['ku:lɔm] n.	库仑
crash [kræʃ] n. & v.	碰撞，坠落，坠毁

215

crystal ['kristəl] n.	水晶，结晶，晶体
crystalline ['krist(ə)lain] adj.	透明的，水晶般的，水晶制的
culmination [ˌkʌlmi'neiʃən] n.	顶点
cumbersome ['kʌmbəsəm] adj.	讨厌的，麻烦的，笨重的
current ['kʌrənt] n.	电流
curvature ['kə:vətʃə] n.	弯曲，曲率
cyan ['saiæn] n.	蓝绿色，青色

D

decode [ˌdi:'kəud] vt.	解码，译解
delay [di'lei] n. & v.	耽搁，延迟
demodulate [di:'mɔdjuleit] v.	解调
demodulation [di:ˌmɔdju'leiʃən] n.	解调
demonstrate ['demənstreit] vt.	示范，证明，论证
depict [di'pikt] vt.	描述，描写
deploy [di'plɔi] vt.	部署
designate ['dezigneit] vt.	指定，指派
detection [di'tekʃən] n.	检波
detector [di'tektə] n.	检波器
device [di'vais] n.	装置，设备
diagram ['daiəgræm] n.	图表，图解
dimension [di'menʃən, dai-] n.	尺寸，尺度，维（数）
diode ['daiəud] n.	二极管
disconnect [ˌdiskə'nekt] vt.	断开
discrete [di'skri:t] adj.	不连续的，离散的
discriminate [dis'krimineit] v.	区别
display ['displei] n.	显示，表现
disruption [dis'rʌpʃn] n.	破裂，毁坏，中断
distortion [dis'tɔ:ʃən] n.	扭曲，变形，曲解，失真
diverse [dai'və:s, di-] adj.	多种多样的

E

earthwards ['ə:θwədz] adj.	向地的
elasticity [ˌelæs'tisəti] n.	弹力，弹性
electrode [i'lektrəud] n.	电极，电焊条
electromagnetic [iˌlektrəumæg'netik] adj.	电磁的
electron [i'lektrɔn] n.	电子

electronic [i'lek'trɔnik] adj.	电子的，电子仪器的
electronics [i'lek'trɔniks] n.	电子学，电学，电子工业
eliminate [i'limineit] vt.	排除，消除，除去
emanate ['eməneit] v.	发出，散发，放射
embed [im'bed] vt.	把……嵌入
embrace [im'breis] v.	拥抱，包括
emerge [i'mə:dʒ] v.	出现
emitter [i'mitə] n.	发射极
encode [in'kəud] vt.	编码
encrypt [in'kript] v.	加密，将……译成密码
energy-efficient adj.	能效高的，高能效的
enormous [i'nɔ:məs] adj.	巨大的
equalizer ['i:kwəlaizə] n.	均衡器，平衡装置
equation [i'kweiʒən] n.	方程式，等式
equator [i'kweitə] n.	赤道
equidistant [,i:kwi'distənt] adj.	距离相等的，等距的
equipment [i'kwipmənt] n.	设备
equivalent [i'kwivələnt] adj.	相等的，相同的，等量的
era ['iərə] n.	时代，纪元，时期
Ethernet ['i:θənet] n.	以太网
evolve [i'vɔlv] v.	发展，进化
expert ['ekspə:t, ek'spə:t] n.	专家，能手
external [ik'stə:nəl] adj.	外部的，客观的
extract [ik'strækt] v.	输出

F

façade [fə'sɑ:d] n.	（建筑物的）正面
facilitate [fə'siliteit] vt.	使容易，促进，帮助
feed [fi:d] v.	供给
femtocell ['femtəusel] n.	家庭基站
fidelity [fi'deləti] n.	逼真，保真度
field ['fi:ld] n.	场
filament ['filəmənt] n.	细丝，灯丝
filter [filtə] n.	滤波器
fire brigade	消防队
flip-flop ['flipflɔp] n.	触发器
flux [flʌks] n.	磁通

217

forward ['fɔ:wəd] adv. 正向
frequency ['fri:kwənsi] n. 频率
function ['fʌŋkʃən] n. & vi. 功能，函数；运行
fundamental [ˌfʌndə'mentl] adj. 基础的，基本的
fuse [fju:z] n. 保险丝，熔丝
fuselage ['fju:zəlɑ:ʒ] n. 机身

G

geostationary ['dʒi:əu'steiʃnri] n. 与地球的相对位置不变的
glacier ['glæsiə] n. 冰川
global ['gləubl] adj. 球形的，全球的，全世界的
glowing ['gləuiŋ] adj. 发光的
grid [grid] n. 格子，网格

H

handheld [hænd,held] adj. 手持式的，便携式的，掌上的
handoff ['hændɔf] n. 切换
hard drive 硬盘
hardware ['hɑ:dweə] n. 计算机硬件
hassle ['hæsl] n. 麻烦事
hierarchy ['haiə,rɑ:ki] n. 层次，层级
hue [hju:] n. 色调，色彩
humidity [hju:'midəti] n. 湿度

I

identification [ai,dentifi'keiʃn] n. 认出，识别
illegal [i'li:gl] adj. 不合法的，非法的
impact ['impækt, im'pækt] n. 碰撞，冲击，影响，效果
impedance [im'pi:dəns] n. 阻抗
implement ['implimənt, 'impliment] n. 工具，器具
implementation [ˌimplimen'teiʃən] n. 成就，贯彻，安装启用
incoming ['inkʌmiŋ] adj. 接踵而来的
inconvenient [ˌinkən'vi:njənt] adj. 不便的，有困难的
incorporate [in'kɔ:pəreit] v. 包含，使并入
inductor [in'dʌktə] n. 感应器，电感
infinite ['infinət] adj. 无限的，无穷的
inherently [in'hiərəntli] adv. 内在地，固有地
instruction [in'strʌkʃn] n. 指令，说明

instrument ['instrəmənt] n.	工具，仪器，器械
instrumentation [ˌinstrəmen'teiʃən] n.	仪器
integrate ['intigreit] v.	使一体化，使集成
integrated ['intigreitid] adj.	综合的，完整的
intensity [in'tensəti] n.	强烈，强度
interact [ˌintər'ækt] v.	交互
intermediate [ˌintə'mi:djət, -dieit] adj. & n.	中间的；媒介
interoperability ['intərˌɔpərə'biləti] n.	互通性，互操作性
interpreter [in'tə:pritə] n.	解释者，翻译器
intersect [ˌintə'sekt] vt. & vi.	横断，横切，贯穿；相交，交叉
intervene [ˌintə'vi:n] n. & vi.	干涉，干预，介入；干涉
intervention [ˌintə'venʃən] n.	干涉
intrusion [in'tru:ʒn] n.	侵扰
inversion [in'və:ʃən] n.	倒置
ionosphere [ai'ɔnəsfiə] n.	电离层
iterative ['itərətiv] adj.	重复的，反复的

J

jukebox ['dʒu:kbɔks] n.	自动点唱机
junction ['dʒʌŋkʃən] n.	连接，接合，交叉点，汇合处

L

label ['leibl] n. & vt.	标签；贴标签于
landscape ['lændskeip] n.	风景，山水画，地形，美化
laptop ['læptɔp] n.	便携式电脑
latency ['leitənsi] n.	潜伏时间，延迟时间
latitude ['lætitju:d] n.	纬度
layout ['leiaut] n.	布局，安排，布置图，规划图
lens [lenz] n.	透镜
light-emitting diode	发光二极管
limitation [ˌlimi'teiʃən] n.	限制，局限性
locate [ləu'keit] v.	位于
longitude ['lɔndʒitju:d] n.	经度
loudspeaker [ˌlaud'spi:kə] n.	扬声器，扩声器，喇叭

M

magnetic [mæg'netik] adj.	磁的，有磁性的，有吸引力的

219

magnitude ['mægnitju:d] n.	大小，数量，巨大，广大
main supply	电源
malfunction [mæl'fʌŋkʃn] n.	故障
matrix ['meitriks] n.	【数】矩阵，模型
measure ['meʒə] v. & n.	测量
mechanism ['mekənizəm] n.	方法，途径，程序，机制，结构
medium ['mi:diəm] n. & adj.	媒体，介质，传导体；中间的，中等的
megabyte ['megəbait] n.	兆字节
meteorite ['mi:tiərait] n.	陨星
microphone ['maikrəfəun] n.	扩音器，麦克风
microsecond ['mikvəusekənd] n.	微秒
microwave ['maikrəweiv] n.	微波
millisecond ['milisekənd] n.	毫秒
moderate ['mɔdərət] adj.	适度的
modify ['mɔdifai] vt.	更改，修改
modular ['mɔdjələ] adj.	模块化的
modulate ['mɔdjuleit] v.	调制
modulation [ˌmɔdju'leiʃən] n.	调制
molecule ['mɔlikju:l] n.	分子
monitor ['mɔnitə] n. & v.	显示器，监视器；监视，监听
monochrome ['mɔnəkrəum] n. & adj.	单色；单色的
monocrystalline [ˌmɔnəu'kristəlain] n. & adj.	单晶体，单晶质；单晶的
molecule ['mɑlikjul] n.	分子，微小颗粒，微粒
motor ['məutə] n.	发动机，电动机
multimeter ['mʌltimi:tə] n.	万用表
multiple ['mʌltipl] adj.	多样的，多重的

N

navigation [ˌnævi'geiʃən] n.	航海
neat [ni:t] adj.	整洁的
needle ['ni:dl] n.	指针
nematic [ni'mætik] adj.	（液晶）【晶体】向列的

O

obstacles ['ɔbstəkl] n.	障碍
obtain [əb'tein] v.	获得
ohmmeter ['əumˌmi:tə] n.	欧姆表

opponent [ə'pəunənt] n. & adj.	对手，反对者；对立的，对抗的
orbit ['ɔ:bit] n. & v.	轨道，常轨；绕轨道而行
orientation [,ɔ:riən'teiʃn] n.	方向，定向
oscilloscope [ə'siləskəup] n.	示波器
output ['autput] n.	输出
overlapping [,əuvə'læpiŋ] adj.	重叠的

P

parameter [pə'ræmitə] n.	参数，参量
passive filter	无源滤波器
period ['piəriəd] n.	时期，周期
phenomenon [fə'nɔminən] n.	现象
phosphor ['fɔsfə] n.	荧光粉
photoconductor [,fəutəukən'dʌktə] n	【物】光电导体，光电导元件
photodiode [,fəutəu'daiəud] n.	光敏二极管，光电二极管
photograph ['fəutəgrɑ:f] n.	照片
phototransistor [,fəutəutræn'zistə] n.	光电晶体管，光敏晶体管
photovoltaics [,fəutəuvɔl'teiiks] n.	光伏，太阳光电，太阳能光电板
pin [pin] n. & v.	大头针，针；拴
pixel ['piks(ə)l] n.	（显示器或电视机图像的）像素（等于 picture element）
plasma ['plæzmə] n.	等离子体
plug [plʌg] vt.& vi.	插上插头
PO number (Purchase Order number)	订单号
polarize ['pəuləraiz] vt.	（使）偏振，（使）极化
polycrystalline [,pɔli'kristəlain] adj.	【晶体】多晶的
positive ['pɔzətiv] adj.	正的
power [pauə] n. & v.	功率；供以动力，激励
precaution [pri'kɔ:ʃn] n.	预防，留心，警戒
precision [pri'siʒən] n.	精确，精密度，精度
primarily ['praimərəli] adv.	首先，起初，主要地，根本上
probe [prəub] n. & v.	探针，探测针；调查
propagation [,prɔpə'geiʃn] n.	传播，传输
protocol ['prəutəkɔl] n.	协议
proportionate [prə'pɔ:ʃənət] adj.	成比例的
prototype ['prəutətaip] n.	原型
provide [prəu'vaid] v.	提供

pseudo ['sju:dəu] adj. 假的，冒充的
pulse [pʌls] n. 脉搏，脉冲

Q

quartz [kwɔrts] n. 石英
quiescent [kwai'esənt] adj. 静止的

R

radar ['reidɑ:] n. 雷达
radiate ['reidieit] v. 发射，辐射
radiation [ˌreidi'eiʃən] n. 辐射，放射，放射线，放射物
reactance [ri'æktəns] n. 电抗
reason ['ri:zn] n. & v. 推理，评理，论证
reception [ri'sepʃən] n. 接收
reconstitute [ˌri:'kɔnstitju:t] vt. 再组成，再构成
rectifier ['rektifaiə] n. 整流器
rectify ['rektifai] v. 整流
redundant [ri'dʌndənt] adj. 多余的
reflect [ri'flekt] vt. & vi. 反射（光、热、声或影像）
reflection [re'flekʃən] n. 反射，映象
refract [ri'frækt] vt. （使）折射，测定……的折射度
refresh cycle 恢复周期，【计】刷新周期　更新周期
regulate ['regjuleit] v. 调节
regulation [ˌregju'leiʃən] n. 规则，调节，校准
reject [ri'dʒekt] v. 去掉，排斥
relative ['relətiv] adj. 相对的
remote sensor 遥感器
replica ['replikə] n. 复制品
replicate ['replikit, 'replikeit] v. 复制
resistor [ri'zistə] n. 电阻器
resonance ['rezənəns] n. 谐振
resonate ['rezəneit] v. 调谐
resonator ['rezəneitə] n. 谐振器
response [ri'spɔns] n. 响应
restrict [ri'strikt] vt. 限制，约束，限定
reverse [ri'və:s] adj. & n. 相反的；反向
rigid ['ridʒid] adj. 刚硬的，刚性的，严格的

risk [risk] n.	风险，危险
robotic [rəu'bɔtik] n.	机器人的
rotational [rəu'teiʃənl] adj.	转动的
router ['ru:tə] n.	路由器

S

satisfactory [ˌsætis'fæktəri] n.	令人满意的
scale [skeil] n.	刻度，衡量，比例，数值范围
scan [skæn] n. & v.	扫描，浏览，审视
seamless ['si:mləs] adj.	无缝的，不停顿的
segment ['segmənt] n.	段，部分
selective [si'lektiv] adj.	有选择的
semiconductor [ˌsemikən'dɔktə] n.	半导体
sensation [sen'seiʃən] n.	感觉，感情，感动
sensor ['sensə] n.	传感器
sequence ['si:kwəns] n.	序列，继起的事，顺序
series ['siəri:z] n.	连续，系列
short-cut ['ʃɔ:tkʌt] n.	捷径
shingle ['ʃiŋgl] n.	瓦板
silicon ['silikən] n.	硅，硅元素
simulation [ˌsimju'leiʃn] n.	仿真，假装，模拟
simultaneously [saiməl'teiniəsli] adv.	同时地
signage ['sainidʒ] n.	引导标示
skip [skip] v.	跳读
solar panel	太阳能板，太阳能电池板
solar radiation	太阳辐射，日光照射
solenoid ['səulənɔid] n.	螺线管
solid ['sɔlid] adj.	固体的
sonar ['səunɑ:] n.	声呐
sophisticated [sə'fistikeitid] adj.	复杂的
spacecraft ['speiskrɑ:ft] n.	宇宙飞船，航天器
span [spæn] v. & n.	横越；跨度，跨距，范围
spark [spɑ:k] vt.	发动，触发
speaker ['spi:kə] n.	扬声器，喇叭
spectral efficiency	频谱效率
stage [steidʒ] n.	级
stereoscopic [ˌsteriəu'skɔpik] adj.	有立体感的

storage ['stɔːridʒ] n.	存储，存储器
store [stɔː] v.	存储
straight [streit] adj.	直的
stuck [stʌk] adj.	动不了的，被卡住的
sturdy ['stəːdi] adj.	结实的，坚固的
substance ['sʌbstəns] n.	物质
suitably ['suːtəbli] adv.	适当地
superimpose [ˌsjuːpərim'pəuz] v.	添加，双重
supersonic [ˌsjuːpə'sɔnik] adj. & n.	超音波的；超声波
surveillance [sə'veiləns] n.	监视
susceptible [sə'septəbl] adj.	易受……影响的
switch [switʃ] n. & vt.	开关，转换；转换
symbol ['simbl] n.	符号，标志，象征
symbology [sim'bɔlədʒi] n.	码制；符号学，符号使用，象征学
synchronise ['siŋkrənaiz] v.	同步（英 synchronize）
synchronization [ˌsiŋkrənai'zeiʃən] n.	同一时刻，同步
syncom ['siŋkɔm] n.	（美国的）同步通信卫星

T

tackle ['tækl] vt.	拦截
target ['tɑːgit] n.	对象，目标
telegraph ['teligrɑːf] n. & v.	电报机，电报；打电报
teleprinter ['teliˌprintə] n.	电传打字机
terrain [tə'rein] n.	地面；地域，地带，【军】地形，地势
terrestrial [ti'restriəl] adj.	地球的，地上的
thermal ['θəːml] adj.	热的，保热的，温热的
thermometer [θə'mɔmitə] n.	温度计，体温计
transaction [træn'zækʃn] n.	交易
transceiver [træn'siːvə] n.	收发器
transfer ['trænsfəː] n.	移动，传递，转移
transform [træns'fɔːm] v.	改变，改观，变换
transformer [træns'fɔːmə] n.	变压器
transistor [træn'sistə] n.	晶体管
transmit [trænz'mit] v.	传输，传送，发送
transparent [træn'spærənt] adj.	透明的，显然的
transponder [træns'pɔndə] n.	应答器
trigger ['trigə] n.	触发器

trilateration [ˌtrailætə'reiʃən] n.	【测】三边测量（术）
tuner ['tju:nə] n.	调谐器
tweak [twi:k] vt.	稍稍调整（机器、系统等）
two-dimensional [ˌtu:di'menʃənəl] adj.	二维的

U

ultimate ['ʌltimət] adj. & n.	最后的，最终的，根本的；最终
ultraviolet [ˌʌltrə'vaiələt] adj. & n.	紫外（线）的；紫外线辐射
unidirectional [ˌju:nidi'rekʃənəl] adj.	单向的，单向性的

V

valve [vælv] n.	电子管，真空管
vegetation [vedʒə'teiʃn] n.	植物
viable ['vaiəbl] adj.	可行的
virtual ['və:tʃuəl] adj.	（计算机）虚拟的
volcano [vɔl'keinəu] n.	火山
volt [vəult] n.	伏特

W

wavelength ['weivleŋθ] n.	波长

APPENDIX II
Abbreviations
常用缩写

A/D (Analog/Digital)	数/模转换
AC (Alternating Current)	交流电
AF (Audio Frequency)	音频
AFC (Automatic Frequency Control)	自动频率控制
AGC (Automatic Gain Control)	自动增益控制
AM (Amplitude Modulated)	调幅
ATM (Asynchronous Transfer Mode)	异步传输模式
B-ISDN (Broadband Integrate Services Digital Network)	
CAI (Computer Aid Instruction)	计算机辅助教学
CCD (Charge Coupled Device)	电荷耦合器件
CCIR (International Radio Consultative Committee)	国际无线电咨询委员会
CCITT (International Telephone and Telegraph Consultative Committee)	
	国际电报电话咨询委员会
CDMA (Code Divide Multiple Address)	码分多址
CD-ROM (Compact disk-ROM)	压缩只读光盘
CRT (Cathode Ray Tube)	阴极射线管
CSPDN (Circuit Switch Public Data Network)	电路分组交换数据网
D/A (Digital/Analog)	数/模转换
DSP (Digital Signal Process)	数字信号处理
DSSS (Direct Sequence Spread Spectrum)	直接序列扩频

DVB (Digital Video Broadcasting)	数字视频广播
DVD (Digital Video Disk)	数字视频光盘
e.m.f (electromotive force)	电动势
FDMA (Frequency Divide Multiple Address)	码分多址
FFT (Fast Flier Transmit)	快速傅里叶变换
FHSS (Frequency-Hopping Spread Spectrum)	跳频扩频
FM (Frequency Modulated)	调频
GPRS (General Packet Radio Service)	通用分组无线业务
GPS (Global Positioning System)	全球定位系统
HD (Hard Disk)	硬盘
HDTV (High Definition TV)	高清晰度电视
HF (High Frequency)	高频
IC (Integrate Circuit)	集成电路
IEEE (Institute of Electrical and Electronics Engineers)	（美国）电机电子工程师协会
IN (Intellect Network)	智能网络
ISDN (Integrate Services Digital Network)	综合业务数字网
ISO (International Standard Organization)	国际标准化组织
ITU (International Telecommunication Union)	国际电信联盟
KCL (Kirchhoff's Current Law)	基尔霍夫电流定理
KVL (Kirchhoff's Voltage Law)	基尔霍夫电压定理
LCD (Liquid Crystal Displayer)	液晶显示器
LED (Light-Emitting Diode)	发光二极管
LF (Loop Filter)	环路滤波器
LTE (Long Term Evolution)	长期演进技术
MIMO (Multi-Input Multi-Output)	多输入多输出技术
MMS (Multimedia Messaging Service)	多媒体短信服务
MOS (Metal-Oxide Semiconductor)	金属氧化物半导体
MPEG (Moving Picture Expert Group)	运动图像专家小组
MW (Medium Wave)	中波
OA (Optical Amplifier)	光放大器
OFDM (Orthogonal Frequency Division Multiplexing)	正交频分复用
OSI (Open System Internet)	开放系统互联
PBCC (Packet Broadcast Control Channel)	分组广播控制信道
PCI (Peripheral Component Interconnect)	周边元件扩展接口
PCM (Pulse Code Modulate)	脉冲编码调制
PD (Phase Detector)	鉴相器
PDA (Personal Digital Assistants)	个人数字助手

PIM (Personal Information Management)	（计算机的）个人信息管理程序
PLL (Phase Locked Loop)	锁相环
PSPDN (Packet Switch Public Data Network)	分组交换公用数据网
PSTN (Public Switching Telephone Network)	公共交换电话网
RF (Radio Frequency)	射频
SC-FDE (Single-Carrier Frequency-Domain-Equalization)	单载波频域均衡
SMS (Short Massage Service)	短消息服务
SNR or S/N (signal-to-noise ratio)	信噪比
SSID (Service Set Identifier)	服务集标识符
SYNC (synchronous communication)	同步通信
3GPP (the 3rd Generation Partnership Project)	第三代合作伙伴项目
TFT (Thin Film Transistor)	薄膜晶体管
TN (Twisted Nematics)	扭曲向列型
UHF (Ultra-High Frequency)	超高频
UMB (Ultra Mobile Broadband)	超移动宽带
USB (Universal Serial Bus)	通用串行总线
VCR (Video Cassette Recorder)	录像机
VGA (Video Graphic Adapter)	可视显示单元
VHF (Very High Frequency)	甚高频
WAP (Wireless Apply Protocol)	无线应用协议
WiMAX (Worldwide Interoperability for Microwave Access)	全球微波互通存取
WMN (Wireless Mesh Network)	无线网状网络
WSN (Wireless Sensor Network)	无线传感器网络